anatomy of
CORE STABILITY

anatomy of
CORE STABILITY

Hollis Lance Liebman

FIREFLY BOOKS

A FIREFLY BOOK

Published by Firefly Books Ltd. 2013

Copyright © 2013 Moseley Road Inc.

All rights reserved. No part of this publication may be reproduced, stored in a retrieval system, or transmitted in any form or by any means, electronic, mechanical, photocopying, recording or otherwise, without the prior written permission of the Publisher.

First printing

Publisher Cataloging-in-Publication Data (U.S.)
A CIP record for this book is available from Library of Congress.

Library and Archives Canada Cataloguing in Publication
A CIP record for this book is available from Library and Archives Canada.

Published in the United States by
Firefly Books (U.S.) Inc.
P.O. Box 1338, Ellicott Station
Buffalo, New York 14205

Published in Canada by
Firefly Books Ltd.
50 Staples Ave., Unit 1
Richmond Hill, Ontario L4B 0A7

Printed in Canada

Anatomy of Core Stability was developed by:
Moseley Road Inc.
123 Main Street
Irvington, New York 10533

Moseley Road Inc.
President: Sean Moore
General Manager: Karen Prince
Art Director: Tina Vaughan
Production director: Adam Moore

Editors: www.sandseditorial.co.uk
Designer Adam Moore

Photography by FineArtsPhotoGroup.com
Models: TJ Fink (tjfink@gmail.com) and Jenna Franciosa

CONTENTS

INTRODUCTION

In our everyday lexicon, words such as "toning," "aerobic," and "lifting" bring to mind a workout at the gym. Other terms—such as "study," "syllabus," and "exam"—suggest we are talking about school. However, certain words transcend the environments with which they are most closely linked. In the past few years, the term "core" has entered the collective consciousness. We hear of toning the core at the gym, firing from the core on the athletic field, getting up from a desk at a sedentary job to engage the core, and strengthening the core at the physical therapist's office. With so many uses and meanings, core is justifiably here to stay.

WHAT EXACTLY IS THE CORE?

The term "core" refers to the muscles in the lower trunk area that work together to provide support and mobility, enabling all bodily movement. The core includes the rectus abdominis, or six-pack, which tenses the abdominal wall by contracting the abdominal muscles. Surrounding your abs are the internal and external obliques, which allow you to bend from side to side and rotate your torso. The Christmas tree–shaped erector spinae is situated behind the abdominals, at the lower back, and is responsible for spinal stabilization and spinal movement. Lastly, the hip flexors act as the foundation of this muscular complex, supporting movement in the pelvic area.

Essentially, the core is the center of the body and is key in terms of performance, functionality, and longevity. Improving the core's aesthetic and mobility necessitates consistent and diligent effort. Getting the most out of your core requires more than just the right diet. Core exercises enable the skeleton, muscles, and joints to work together properly, as well as offering a combination of strengthening, stretching, balancing, realigning, and fat loss that nutrition alone would not achieve.

By maintaining a strong core, you also lend optimal support to ancillary (assisting) muscles. Indeed, the core is so central to your body's total movement that it is called upon when firing each and every muscle. When you are in the middle of a squat, for example, your core is engaged to maintain the integrity of the vertical movement; when pressing dumbbells overhead, the core keeps the body straight on, as opposed to curved. Have you ever worked your triceps and discovered later that your midsection was quite sore? That's your core at work.

Even away from the gym, real-life everyday tasks are made possible only by the core stabilizing your body and offering support to other firing muscles. Searching the cupboard for something to eat? You are using your core. Mowing the lawn, changing a diaper, putting groceries away, and myriad other mundane chores

The Sumo Squat engages the core to perform the correct vertical movement.

employ not just your body musculature but core stabilization, too.

Unfortunately, artificial supports such as chair backs have, for years, done the body's work for us, and our cores have suffered as a result. In particular, those with sedentary vocations ought to service their cores regularly. Core training is not just for athletes; it is for everyone—from the golfer looking to improve his or her game, to the office worker sitting at a desk all day and complaining of back pain. And core training will benefit you, the busy person with far greater responsibilities than ever before and less time in which to work out.

CORE TRAINING VS. CORE STABILITY

With so much talk about core, the terms "core training" and "core stability" are often used interchangeably. However, they are actually two very different things. In core training, the muscles function as a unified whole rather than in isolation, as they would, for example, in most weight-resistance programs, which are targeted more toward particular muscle

A well-maintained core will help you do whatever activities you desire, with minimal difficulty.

groups—chest and biceps, say. Core exercises are movements and physical positions that target the core directly, and it is possible to use core training—much like weight training—to enhance your midsection's muscular definition and compactness throughout. In this way, core training is often the goal in and of itself for those wishing to create a more defined, sleeker midsection, such as models or bodybuilders, who wish to show off great muscle definition. Core-stability exercises, by contrast, are those movements and physical positions that help build a strong core that is ready to do whatever you ask of it. The core is employed in all movement and is nearly impossible to exclude while working out. Were you to press a weight overhead, you would find that your core acts as a stabilizer, allowing the deltoids and triceps to complete the given task. Were it not for even the most rudimentary core stability, the torso would simply buckle instead of remaining upright and erect during the movement. The function of the abdominals is to assist the spine, and stability exercises improve your ability to do this, while also working the visible abdominal muscles.

9

INTRODUCTION

CORE STABILITY VS. INSTABILITY?

If core stability were the result of spinal support for real-world motion, then instability would be the surrounding joint structures' lack of support against the spine, making movement difficult. Voluntary core movement is largely the result of primary muscles or muscle groups such as the rectus abdominis and erector spinae, but it is also helped by smaller muscles such as the transversus abdominis. In a healthy individual, the stabilizer muscles work automatically, but where there is an injury—such as a sprain or disc herniation—the spine is not supported properly. In such cases, the primary movers take over and are forced to do the work of the collective; this can lead to instability.

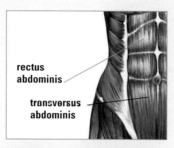

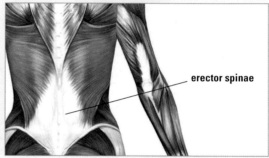

Core stability relies heavily on the rectus abdominis, transversus abdominis, and erector spinae.

rectus abdominis

transversus abdominis

erector spinae

You can liken optimal core stability to greater functionality or real-world performance, rather than pure aesthetics or the shape and clarity of abdominal musculature. It is possible to have very good core stability (performance and support) and yet appear soft or fleshy throughout the midsection. However, in stabilizing the core through exercise, you are also working on making the abdominals more visible. Conversely, just because somebody is physically strong, it does not necessarily make them stable. Strength is a factor of core stability, but it is not the main function.

As we age, the blind pursuit of aesthetics tends to diminish somewhat, and the desire to be healthy far exceeds the will to look good. This book can help you sculpt a leaner, more pleasing, stronger waistline, which in turn will provide support for the ever-aging spine, helping you feel more youthful.

STATIC VS. DYNAMIC EXERCISES

In working and strengthening the core, two different exercise forms must be employed in order to achieve the best results: static and dynamic. Static exercise results in greater strength, flexibility, and mobility, whereas dynamic exercise leads to improved blood circulation, strength, and endurance.

Although people often think of abdominal exercise when considering core work, truly functional training is far more beneficial—that is, training that stimulates improvement in the body's ability to complete everyday tasks. It is important to learn how to engage your core in various positions, as well as during activity, to provide maximum stability for your spine.

USING THIS BOOK

Anatomy of Core Stability begins with a selection of routines that can be considered warm-ups or stretches. These are followed by a rich selection of static and dynamic exercises, and the book concludes with actual workout sequences. It is important to ease into the exercises outlined. Since the primary goal is core stabilization, strength and the explosiveness of the movements are secondary to mastering correct form and execution, as well as the assistance of ancillary muscles and, above all, proper spinal support.

Several mainstays to keep in mind when working the core include correct breathing, speed, and movement for the exercises. With a diligent and firm command of these three elements, it is possible to tap into the muscles efficiently and effectively. It is for this reason that endless sets and repetitions are neither necessary nor advised. You need merely undertake a few calculated sets to achieve a deep muscular burn.

Breathing should be natural and steady. A deep inhale should accompany the negative, or stretched, part of the movement (think of pulling back the arrow on a bow before launch), followed by a full exhale on the positive, or extended, portion (releasing the arrow). It is of paramount importance never to hold your breath during exercise, particularly during the static exercises, since this could ultimately prove fatal. Be conscious of your breathing, but also remain attuned to the exercise at hand.

Exercise speed should be based on a slow or controlled negative portion of the movement, followed by an explosive positive. I advise 5-second counted reps. These should be neither very fast nor super-slow; rather, they should be at a natural pace that can be maintained throughout the entire set.

The key to effective exercise is to focus fully on each and every repetition. Those who claim to complete 1,000 sit-ups would, in truth, be lucky to have executed 100 properly, since the neck and lower back tend to be called into play, as well as speed and momentum, when performing that many— and, of course, they should not. Although the core calls for a working together of both primary and ancillary muscles, core stability is at its best when each muscle performs its job as part of a collective.

Basic exercises such as the Push-Up or Sit-Up are only beneficial when performed correctly. Think quality over quantity.

FULL-BODY ANATOMY

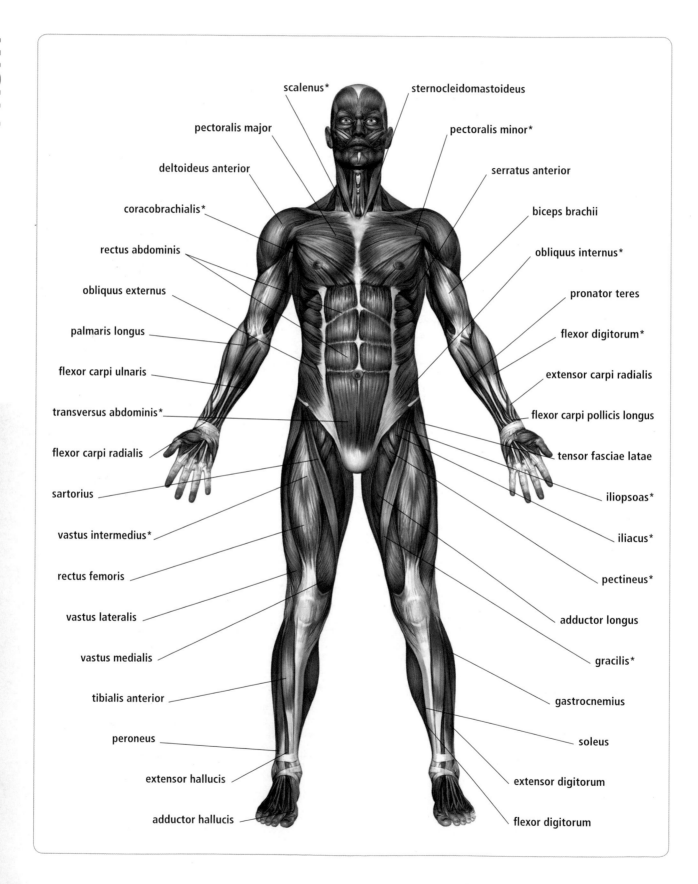

scalenus*

sternocleidomastoideus

pectoralis major

pectoralis minor*

deltoideus anterior

serratus anterior

coracobrachialis*

biceps brachii

rectus abdominis

obliquus internus*

obliquus externus

pronator teres

palmaris longus

flexor digitorum*

flexor carpi ulnaris

extensor carpi radialis

transversus abdominis*

flexor carpi pollicis longus

flexor carpi radialis

tensor fasciae latae

sartorius

iliopsoas*

vastus intermedius*

iliacus*

rectus femoris

pectineus*

vastus lateralis

adductor longus

vastus medialis

gracilis*

tibialis anterior

gastrocnemius

peroneus

soleus

extensor hallucis

extensor digitorum

adductor hallucis

flexor digitorum

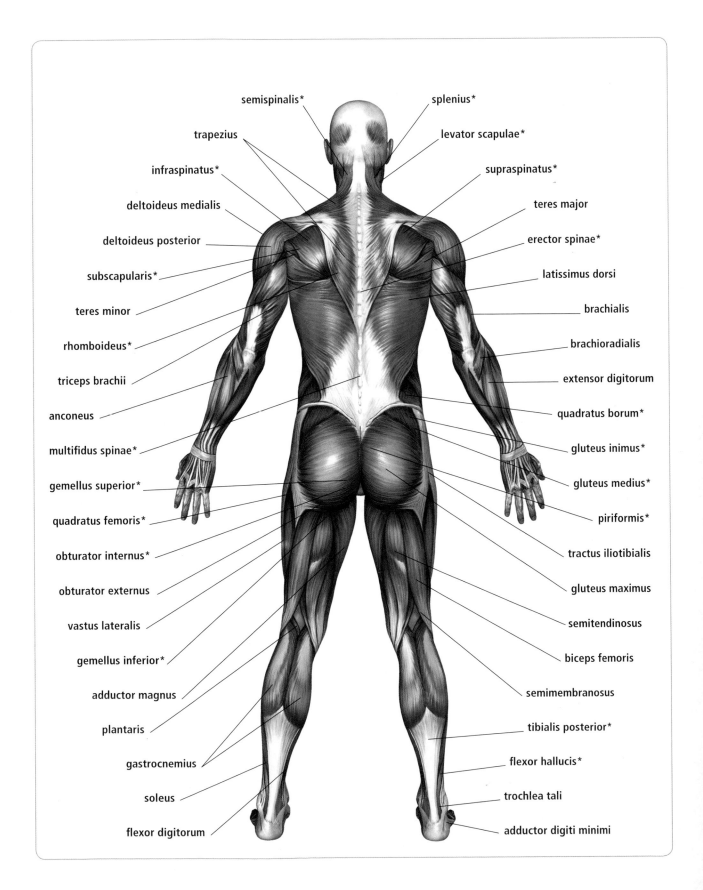

semispinalis*

splenius*

trapezius

levator scapulae*

infraspinatus*

supraspinatus*

deltoideus medialis

teres major

deltoideus posterior

erector spinae*

subscapularis*

latissimus dorsi

teres minor

brachialis

rhomboideus*

brachioradialis

triceps brachii

extensor digitorum

anconeus

quadratus borum*

multifidus spinae*

gluteus inimus*

gemellus superior*

gluteus medius*

quadratus femoris*

piriformis*

obturator internus*

tractus iliotibialis

obturator externus

gluteus maximus

vastus lateralis

semitendinosus

gemellus inferior*

biceps femoris

adductor magnus

semimembranosus

plantaris

tibialis posterior*

gastrocnemius

flexor hallucis*

soleus

trochlea tali

flexor digitorum

adductor digiti minimi

WARM-UP & STRETCHING EXERCISES

In any effective workout program, a proper warm-up is essential to

help prevent injury and to provide adequate blood flow to the working

muscles, making them more pliable for the intense contractions that core

work sometimes requires. It is always best, even prior to stretching,

to raise your body temperature slightly and to elevate your heart rate so

that the muscle being stretched won't tear but, instead, will be ready

for exercise. A good 5 to 10 minutes of light cardiovascular activity—

for example, on a stationary bike on a low setting—is sufficient.

SWISS BALL ABDOMINAL STRETCH

❶ Begin on your back on a Swiss ball, with your feet shoulder-width apart and your arms extended behind your head.

❷ Reach your arms backward until your hands touch the floor.

DO IT RIGHT
Keep your torso planted on the ball.

AVOID
Overextending your pelvic raise.

❸ While keeping your lower back on the ball, lower your hips and stretch your abdominals toward the ceiling. Hold for 30 seconds, relax, and repeat for another 30 seconds.

LEVEL
• Beginner

TIME
• 30-second hold; 1-minute completion time

BENEFITS
• Stretches the rectus abdominis

TARGET AREAS
Primary emphasis is on the rectus abdominis.

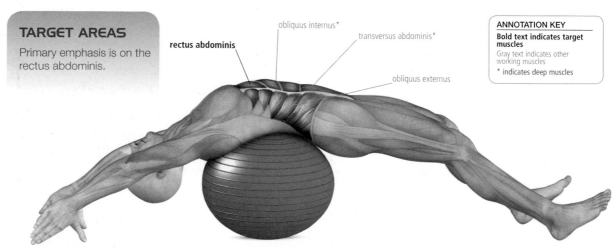

rectus abdominis

obliquus internus*

transversus abdominis*

obliquus externus

ANNOTATION KEY
Bold text indicates target muscles
Gray text indicates other working muscles
* indicates deep muscles

SIDE STRETCH

1 Begin standing with your right hand on your hip and your left arm over your head. Reach your left hand over toward the right side, leaning your torso in the same direction. Hold for 30 seconds.

DO IT RIGHT	AVOID
Keep your torso straight on.	Bending forward or backward at the waist.

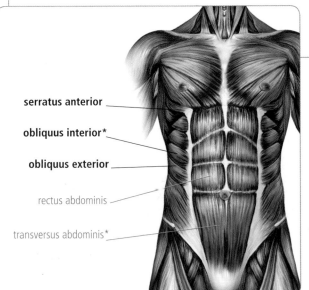

2 Relax, and repeat for another 30 seconds, then switch sides.

LEVEL
• Beginner

TIME
• 30 second hold; 2-minute completion time

BENEFITS
• Stretches the upper back and core

TARGET AREAS

Primary emphasis is on the upper-back, serratus, oblique, and intercostal muscles.

ANNOTATION KEY
Bold text indicates target muscles
Gray text indicates other working muscles
* indicates deep muscles

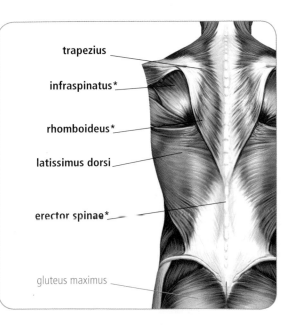

trapezius

infraspinatus*

rhomboideus*

latissimus dorsi

erector spinae*

gluteus maximus

serratus anterior

obliquus interior*

obliquus exterior

rectus abdominis

transversus abdominis*

17

SIDE BENDS

1 Stand upright with your hands resting on your hips. Bend slowly to the right, sliding your right hand down your thigh at the same time, and then return to the vertical position.

DO IT RIGHT
Keep your torso straight on.

AVOID
Bending forward or backward at the waist.

2 Perform ten repetitions, then switch to the other side.

LEVEL
• Beginner

TIME
• 1-minute completion time

BENEFITS
• Stretches the core

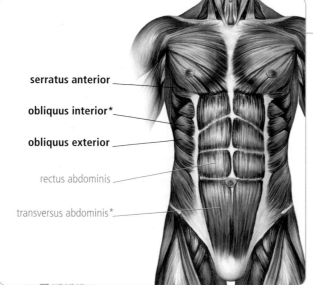

serratus anterior

obliquus interior*

obliquus exterior

rectus abdominis

transversus abdominis*

TARGET AREAS
Primary emphasis is on the serratus, oblique, and intercostal muscles.

ANNOTATION KEY
Bold text indicates target muscles
Gray text indicates other working muscles
* indicates deep muscles

SEATED SPINAL STRETCH

1 Sit on the floor with your right leg stretched out in front of you and your left leg bent at the knee and placed across the right leg, with the foot on the ground. Keep your left hand on the ground for support and the right draped over your left leg.

2 Rotate your torso to the left. Hold for 30 seconds, repeat, and then switch sides.

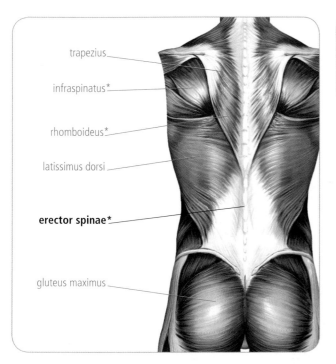

trapezius

infraspinatus*

rhomboideus*

latissimus dorsi

erector spinae*

gluteus maximus

DO IT RIGHT

Keep your back straight at all times.

AVOID

Excessively rotating your torso during the stretch.

TARGET AREAS

Primary emphasis is on the erector spinae.

ANNOTATION KEY

Bold text indicates target muscles

Gray text indicates other working muscles

* indicates deep muscles

LEVEL
• Beginner

TIME
• 30-second hold; 2-minute completion time

BENEFITS
• Leads to increased mobility of the spine

RESTRICTIONS
• Those with lower-back problems should avoid this exercise.

COBRA STRETCH

① Lie facedown with your arms bent, your elbows in, and your palms on the ground.

DO IT RIGHT
Keep your arms close to your sides.

AVOID
An excessive upward swing.

② Lift your upper body until your arms are at full length. Complete three repetitions of 15 seconds each.

LEVEL
• Intermediate

TIME
• 45-second completion time

BENEFITS
• Helps loosen the spinal joints

RESTRICTIONS
• Those with lower-back problems should avoid this exercise.

erector spinae*

obliquus externus

quadratus lumborum

ANNOTATION KEY
Bold text indicates target muscles
Gray text indicates other working muscles
* indicates deep muscles

TARGET AREAS
Primary emphasis is on the erector spinae.

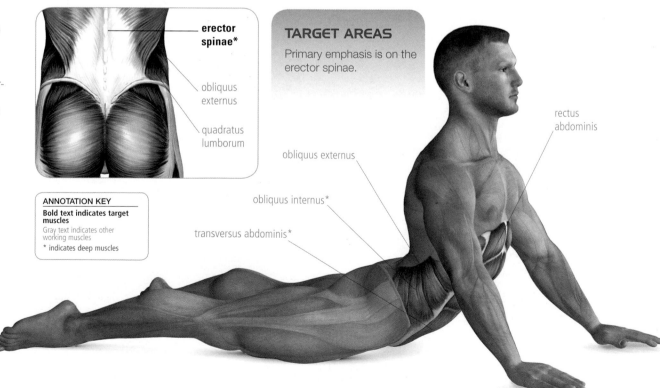

rectus abdominis

obliquus externus

obliquus internus*

transversus abdominis*

ILIOTIBIAL BAND STRETCH

1 Start in a standing position, and cross your left foot behind the right ankle. Raise your arms over your head, and lean forward until you are as close to the floor with your fingertips as you can go.

2 Hold for 20 seconds and repeat, then switch legs and repeat the entire stretch.

DO IT RIGHT

Be sure to ease into the movement.

AVOID

Overextending your legs.

LEVEL
• Intermediate

TIME
• 20-second hold; 90-second completion time

BENEFITS
• Increases range of hip movement

tractus iliotibialis

gluteus maximus

vastus lateralis

semitendinosus

biceps femoris

semimembranosus

TARGET AREAS

Primary emphasis is on the tractus iliotibialis.

ANNOTATION KEY
Bold text indicates target muscles
Gray text indicates other working muscles
* indicates deep muscles

ADDUCTOR STRETCH

1 Stand tall with your feet wide apart.

2 Bend your right leg while lowering your torso to the ground. Rest your hands on your thighs while you feel the deep stretch inside the left thigh.

3 Hold for 30 seconds, relax, and repeat for another 30 seconds, then switch legs and repeat the exercise.

DO IT RIGHT
Keep your torso straight on.

AVOID
Overstretching your extended leg.

LEVEL
• Beginner

TIME
• 30-second hold; 2-minute completion time

BENEFITS
• Stretches the adductors

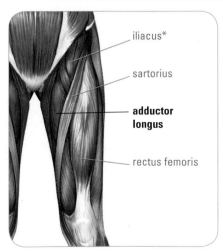

iliacus*

sartorius

adductor longus

rectus femoris

TARGET AREAS
Primary emphasis is on the adductor muscles.

ANNOTATION KEY
Bold text indicates target muscles
Gray text indicates other working muscles
* indicates deep muscles

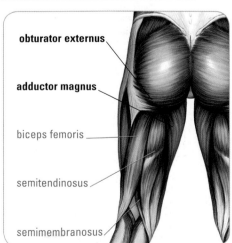

obturator externus

adductor magnus

biceps femoris

semitendinosus

semimembranosus

HIP FLEXOR STRETCH

1 Kneel on your left leg with your right leg bent in front of you. Place your right hand on your right leg, and let your left arm hang down to your thigh or place it on your hip.

DO IT RIGHT

Keep your front knee directly over your foot for support.

AVOID

Excessively straining your front thigh.

2 Shift your weight to your right thigh, feeling a deep stretch within, while keeping your chest out and your back flat. Hold for 30 seconds, relax, and repeat for another 30 seconds, then switch legs.

LEVEL
• Beginner

TIME
• 30-second hold; 2-minute completion time

BENEFITS
• Increases your ability to lift the knees and bend at the waist

iliopsoas*

iliacus*

pectineus*

sartorius

TARGET AREAS

Primary emphasis is on the hip flexors.

ANNOTATION KEY
Bold text indicates target muscles
Gray text indicates other working muscles
* indicates deep muscles

PIRIFORMIS STRETCH

1 Lie on your back with your left leg bent and your right ankle crossed over your left knee. Use your hands to grab the back of the left thigh, close to the knee, and gently pull it toward your right shoulder.

DO IT RIGHT
Keep your back pressed to the ground.

AVOID
Excessively pulling on or straining the knee.

2 Hold for 30 seconds, relax, and repeat for another 30 seconds, then switch sides.

LEVEL
• Beginner

TIME
• 30-second hold; 2-minute completion time

BENEFITS
• Stretches the gluteal and hip regions

TARGET AREAS
Primary emphasis is on the gluteal muscles and hips.

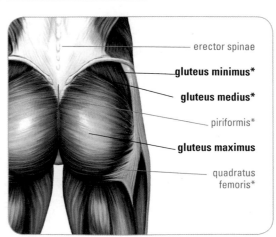

erector spinae

gluteus minimus*

gluteus medius*

piriformis*

gluteus maximus

quadratus femoris*

ANNOTATION KEY
Bold text indicates target muscles
Gray text indicates other working muscles
* indicates deep muscles

LUMBAR STRETCH

1 Lie on your back with your legs bent to a 90-degree angle and your arms extended outward.

DO IT RIGHT

Keep your back pressed flat against the floor.

AVOID

Jerking your leg hard across your side.

2 Gently pull your knees over to your left side until the bottom knee almost touches the floor. Hold for 30 seconds, repeat, and then switch sides.

LEVEL
• Beginner

TIME
• 30-second hold; 2-minute completion time

BENEFITS
• Helps keep your spine flexible

RESTRICTIONS
• Those with lower-back problems should avoid this exercise.

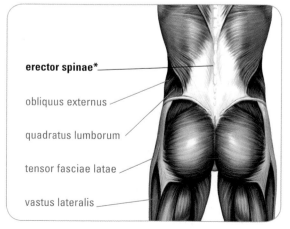

erector spinae*

obliquus externus

quadratus lumborum

tensor fasciae latae

vastus lateralis

TARGET AREAS

Primary emphasis is on the erector spinae.

ANNOTATION KEY
Bold text indicates target muscles
Gray text indicates other working muscles
* indicates deep muscles

STATIC EXERCISES

Static exercise, also known as isometrics, greatly exerts muscles without requiring movement of the joints. Pushing on an immovable wall is an example of static exercise. Stretching exercises can be considered static because the posture is held. The primary function of static exercise is to stabilize the spine, which should not move. During the exercises, the length of the muscle used does not change, and there is no visible movement at the joint. Do not hold your breath during isometrics, because static exercise can significantly raise blood pressure, and holding your breath could prove fatal in extreme circumstances. Those with cardiovascular disease or hypertension should approach this form of exercise with caution.

STANDING STABILITY

1 Begin in a standing position with your left foot on a foam block and your right leg bent at a 90-degree angle. Hold your arms fully extended at 90 degrees to your body, parallel to the ground.

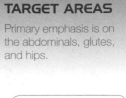

2 Hold for 30 seconds, repeat, and then switch sides.

LEVEL
• Intermediate

TIME
• 2-minute completion time

BENEFITS
• Helps stabilize the body as a whole

RESTRICTIONS
• Those with lower-back problems should avoid this exercise.

MODIFICATION

To push yourself a little harder, try performing this exercise with your eyes closed.

gluteus medius*

gastrocnemius

soleus

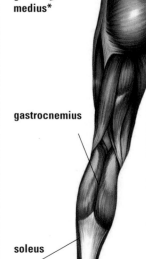

rectus abdominis

transversus abdominis*

iliopsoas*

iliacus*

TARGET AREAS

Primary emphasis is on the abdominals, glutes, and hips.

ANNOTATION KEY

Bold text indicates target muscles
Gray text indicates other working muscles
* indicates deep muscles

STANDING EXTENSION

❶ Begin in a standing position with your hands on your hips.

DO IT RIGHT
Keep your torso and abdominals contracted.

❷ While keeping your abdominals braced, lean back as far as is comfortable, and hold. Repeat 10 times.

AVOID
Slouching your shoulders.

LEVEL
• Intermediate

TIME
• 1-minute completion time

BENEFITS
• Helps stabilize the body as a whole

RESTRICTIONS
• Those with lower-back problems should avoid this exercise.

TARGET AREAS
Primary emphasis is on the rectus abdominis and the erector spinae.

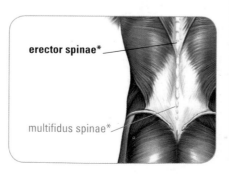

erector spinae*

multifidus spinae*

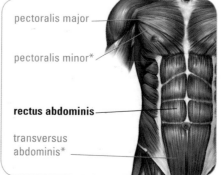

pectoralis major

pectoralis minor*

rectus abdominis

transversus abdominis*

ANNOTATION KEY
Bold text indicates target muscles
Gray text indicates other working muscles
* indicates deep muscles

STANDING ABDOMINAL BRACING

1 Stand with your legs slightly apart and your arms folded in front of you.

DO IT RIGHT	AVOID
Be sure to exhale while contracting.	Performing the exercise at an excessive speed.

2 Rock your pelvis forward slightly and then backward, holding for three repetitions of 5-second counts.

LEVEL
- Beginner

TIME
- 15-second completion time

BENEFITS
- Helps loosen the core area

RESTRICTIONS
- Those with lower-back problems should avoid this exercise.

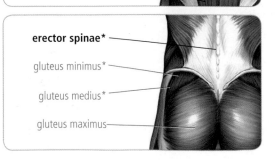

rectus abdominis

transversus abdominis*

erector spinae*

gluteus minimus*

gluteus medius*

gluteus maximus

TARGET AREAS

Primary emphasis is on the abdominals and lower back.

SEATED PELVIC TILT

DO IT RIGHT

Be sure to exhale while contracting.

AVOID

Performing the exercise at an excessive speed.

1 Sit on a Swiss ball with your hands resting on your thighs.

2 While contracting the abdominals, gently rock your torso front to back and side to side for five repetitions of 5-second counts.

TARGET AREAS

Primary emphasis is on freeing up the spine.

LEVEL
• Beginner

TIME
• 30-second hold; 2-minute completion time

BENEFITS
• Leads to increased mobility of the spine

RESTRICTIONS
• Those with lower-back problems should avoid this exercise.

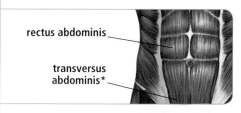

rectus abdominis

transversus abdominis*

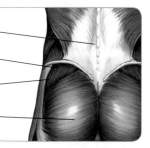

erector spinae*

gluteus minimus*

gluteus medius*

gluteus maximus

ANNOTATION KEY
Bold text indicates target muscles
Gray text indicates other working muscles
* indicates deep muscles

CHAIR POSE

1 Start by standing in an upright position.

DO IT RIGHT

Keep your abdominals contracted throughout the exercise.

AVOID

Avoid arching your back excessively.

2 Raise your arms over your head, bend your knees, and extend your upper body forward to an approximately 45-degree angle.

LEVEL
• Beginner

TIME
• 1-minute completion time

BENEFITS
• Helps stabilize the body as a whole

RESTRICTIONS
• Those with lower-back problems should avoid this exercise.

3 Keep your feet flat, and push through your heels. Hold for 30–60 seconds.

TARGET AREAS

Primary emphasis is on the lower back, quadriceps, and calves, with secondary assistance from the triceps and deltoids.

ANNOTATION KEY

Bold text indicates target muscles

Gray text indicates other working muscles

* indicates deep muscles

STATIC

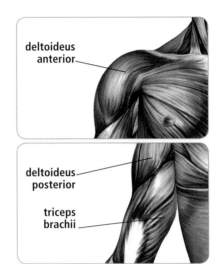

deltoideus anterior

deltoideus posterior

triceps brachii

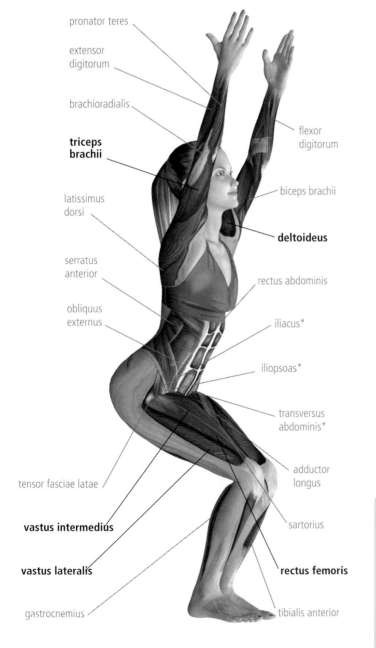

pronator teres

extensor digitorum

brachioradialis

triceps brachii

latissimus dorsi

serratus anterior

obliquus externus

tensor fasciae latae

vastus intermedius

vastus lateralis

gastrocnemius

flexor digitorum

biceps brachii

deltoideus

rectus abdominis

iliacus*

iliopsoas*

transversus abdominis*

adductor longus

sartorius

rectus femoris

tibialis anterior

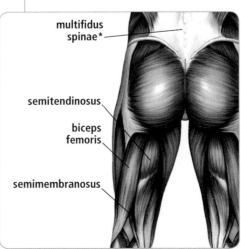

multifidus spinae*

semitendinosus

biceps femoris

semimembranosus

33

HAND-TO-TOE LIFT

STATIC

1 Stand with your right hand on your hip and your weight shifted to the right foot.

2 Raise your left knee toward your chest, and take hold of your left foot with your left hand.

DO IT RIGHT

Keep your hips straight on and squared up.

AVOID

Bouncing around on the foot.

LEVEL
- Intermediate

TIME
- 2-minute completion time

BENEFITS
- Will increase abdominal and leg stability

RESTRICTIONS
- Those with lower-back problems should avoid this exercise.

MODIFICATION

For a harder challenge, add this step before lowering your leg. Swing your left leg out to the side, still holding your toes. Breathe steadily, and hold for about 5 seconds.

3 Extend the left leg out in front of you, keeping hold of the toes with your fingers. Maintain the position for 10 seconds, and then lower the leg. Perform five repetitions per leg.

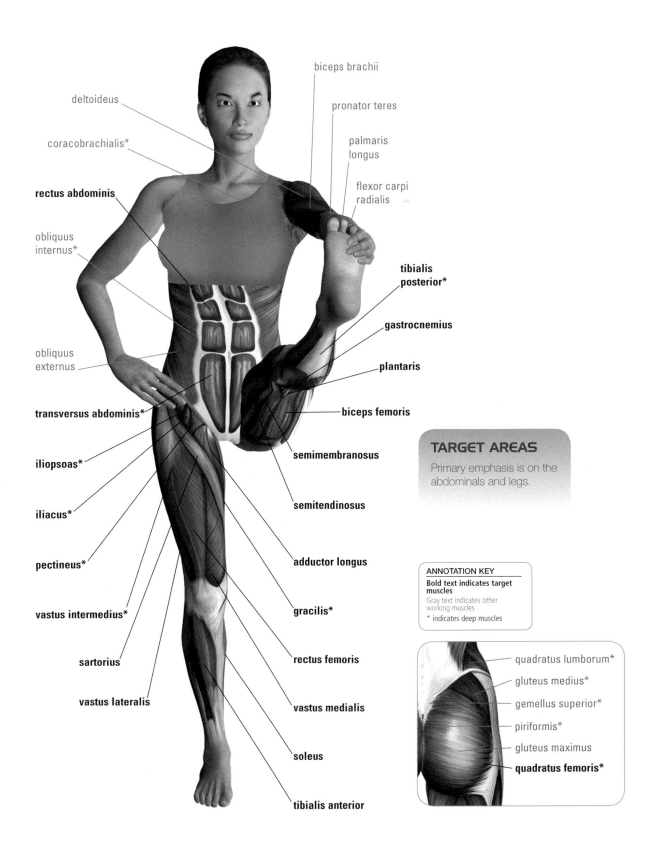

deltoideus

coracobrachialis*

rectus abdominis

obliquus
internus*

obliquus
externus

transversus abdominis*

iliopsoas*

iliacus*

pectineus*

vastus intermedius*

sartorius

vastus lateralis

biceps brachii

pronator teres

palmaris
longus

flexor carpi
radialis

**tibialis
posterior***

gastrocnemius

plantaris

biceps femoris

semimembranosus

semitendinosus

adductor longus

gracilis*

rectus femoris

vastus medialis

soleus

tibialis anterior

TARGET AREAS

Primary emphasis is on the
abdominals and legs.

ANNOTATION KEY
**Bold text indicates target
muscles**
Gray text indicates other
working muscles
* indicates deep muscles

quadratus lumborum*

gluteus medius*

gemellus superior*

piriformis*

gluteus maximus

quadratus femoris*

SITTING BALANCE

1 Sit on a Swiss ball, with your hands resting on the ball at your sides.

DO IT RIGHT	AVOID
Keep your core contracted.	Slouching forward.

2 Raise your right leg parallel to the ground, and hold for 5 seconds.

3 Repeat with the left leg. Perform five repetitions per leg.

LEVEL
• Beginner

TIME
• 1-minute completion time

BENEFITS
• Helps strengthen and stabilize the abdominals

RESTRICTIONS
• Those with lower-back problems should avoid this exercise.

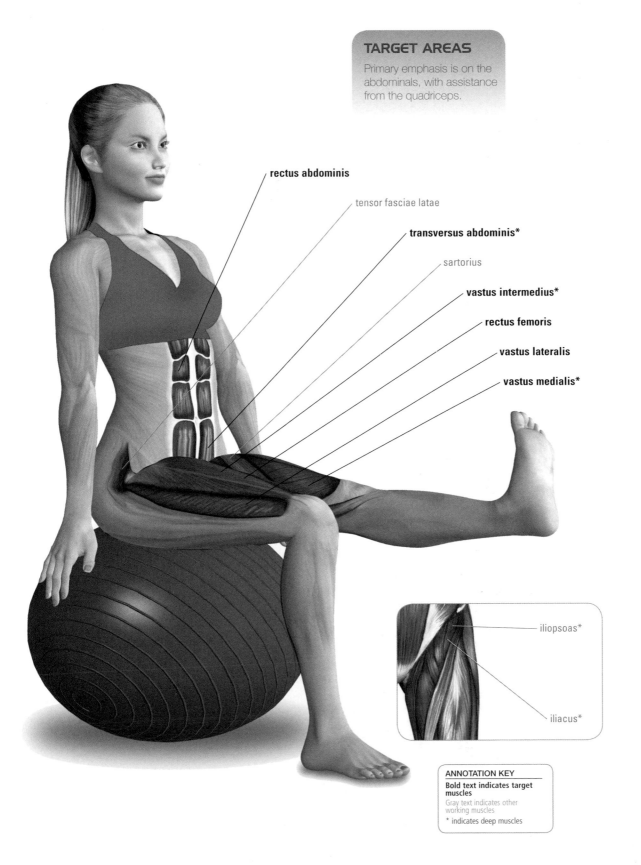

TARGET AREAS
Primary emphasis is on the abdominals, with assistance from the quadriceps.

rectus abdominis

tensor fasciae latae

transversus abdominis*

sartorius

vastus intermedius*

rectus femoris

vastus lateralis

vastus medialis*

iliopsoas*

iliacus*

ANNOTATION KEY
Bold text indicates target muscles
Gray text indicates other working muscles
* indicates deep muscles

37

THIGH ROCK-BACK

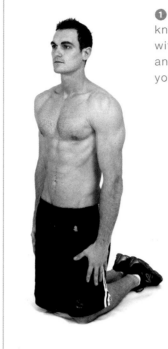

1 Begin in a kneeling position, with a straight back and your arms at your sides.

DO IT RIGHT
Maintain a straight line with your torso.

AVOID
Leaning back too far.

2 Lean back while keeping your body in a straight line and your abdominals contracted.

3 While still leaning back, flex your glute muscles, then slowly return to the starting position. Complete 10 repetitions.

LEVEL
• Advanced

TIME
• 1-minute completion time

BENEFITS
• Will improve abdominal and thigh strength

RESTRICTIONS
• Those with lower-back problems should avoid this exercise.

TARGET AREAS

Primary emphasis is on the abdominals and quadriceps.

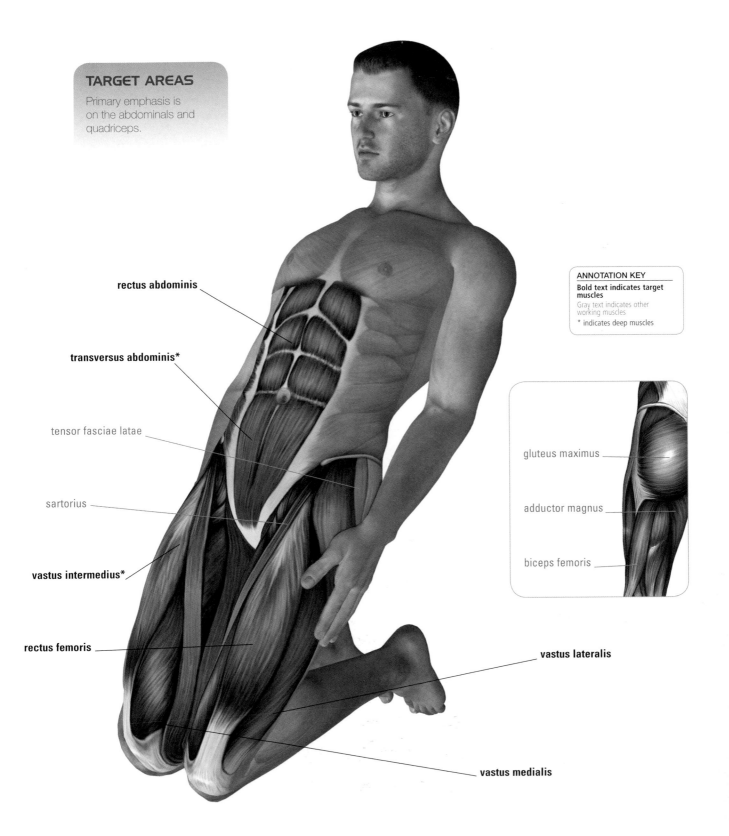

rectus abdominis

transversus abdominis*

tensor fasciae latae

sartorius

vastus intermedius*

rectus femoris

vastus lateralis

vastus medialis

ANNOTATION KEY

Bold text indicates target muscles

Gray text indicates other working muscles

* indicates deep muscles

gluteus maximus

adductor magnus

biceps femoris

PLANK

❶ Position yourself on all fours, then plant your forearms on the floor parallel to one another, with 90-degree bends at the elbows.

DO IT RIGHT

Keep your abdominal muscles tight and your body in a straight line.

AVOID

Bridging too high, since this can take stress off the working muscles.

❷ Raise your knees off the ground, and lengthen your legs until they are in line with your body. Hold for 30 seconds (building up to 120 seconds).

LEVEL
• Beginner to intermediate

TIME
• Beginner: 30-second completion time; Intermediate: 2-minute completion time

BENEFITS
• Increases the ability to support your own body weight

RESTRICTIONS
• Pregnant women, though not restricted, should exercise caution when performing the Plank.

TARGET AREAS

Primary emphasis is on the rectus abdominis and erector spinae.

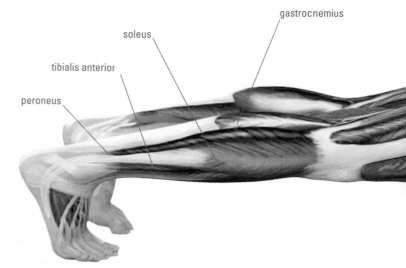

gastrocnemius

soleus

tibialis anterior

peroneus

40

MODIFICATION

This variant offers a more challenging exercise. Instead of resting on your forearms, extend your arms fully when on all fours, then continue with step 2 (opposite).

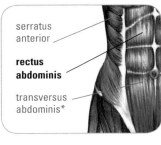

serratus anterior

rectus abdominis

transversus abdominis*

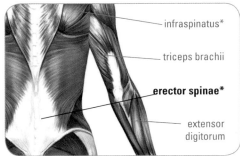

infraspinatus*

triceps brachii

erector spinae*

extensor digitorum

ANNOTATION KEY

Bold text indicates target muscles

Gray text indicates other working muscles

* indicates deep muscles

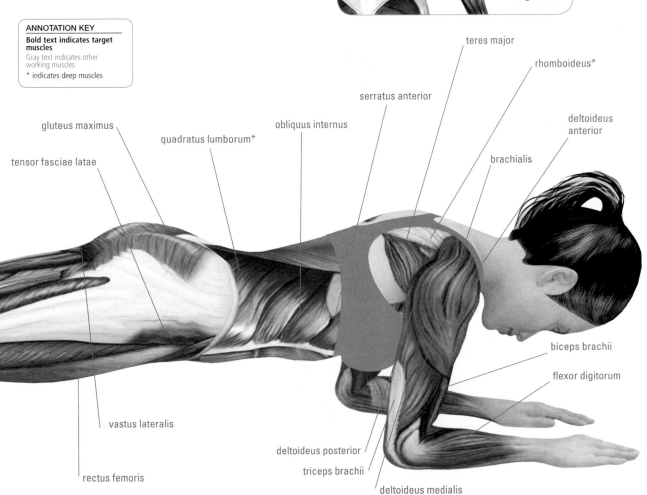

teres major

rhomboideus*

serratus anterior

deltoideus anterior

gluteus maximus

quadratus lumborum*

obliquus internus

brachialis

tensor fasciae latae

biceps brachii

flexor digitorum

vastus lateralis

deltoideus posterior

triceps brachii

rectus femoris

deltoideus medialis

SIDE PLANK

❶ Lie on your right side with your legs straight and parallel to one another.

❷ Bend your right arm to a 90-degree angle with your knuckles facing forward. Rest your left arm along your left hip.

DO IT RIGHT

Push evenly from both your forearm and hips.

AVOID

Placing too much strain on the shoulders.

LEVEL
• Advanced

TIME
• 2-minute completion time

BENEFITS
• Increases isometric strength for trunk stabilization

RESTRICTIONS
• Sufferers of chronic lower-back pain should be cautious when attempting this exercise.

❸ Push through your right forearm while raising your hips off the ground until your body is one straight line.

❹ Hold for 30 seconds (working up to 1 full minute), then switch to your left side and repeat the exercise.

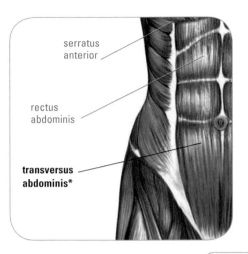

serratus anterior

rectus abdominis

transversus abdominis*

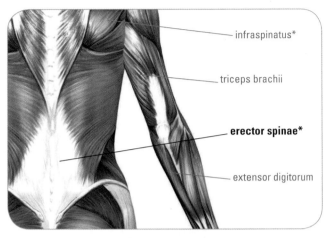

infraspinatus*

triceps brachii

erector spinae*

extensor digitorum

ANNOTATION KEY
Bold text indicates target muscles
Gray text indicates other working muscles
* indicates deep muscles

TARGET AREAS

Primary emphasis is on the transversus abdominis, erector spinae, and deltoids.

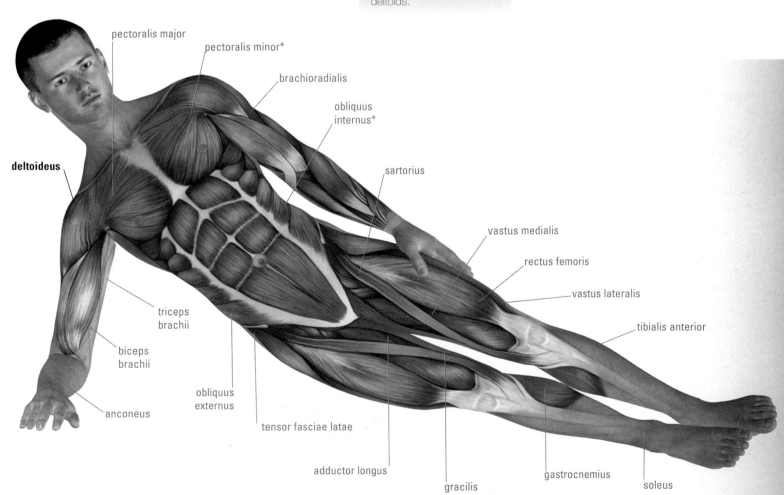

pectoralis major

pectoralis minor*

brachioradialis

obliquus internus*

sartorius

deltoideus

vastus medialis

rectus femoris

vastus lateralis

tibialis anterior

triceps brachii

biceps brachii

anconeus

obliquus externus

tensor fasciae latae

adductor longus

gracilis

gastrocnemius

soleus

FRONT PLANK

① Sit with your legs extended in front of you and your arms directly behind you, with your fingers pointing straight ahead.

DO IT RIGHT

DO IT RIGHT

Keep your pelvis elevated for the duration of the exercise.

AVOID

Letting your shoulders slouch backward.

② Push through your palms and raise your hips and glutes off the ground until your body forms a straight line from the shoulders down.

LEVEL
• Intermediate

TIME
• 1-minute completion time

BENEFITS
• Increases the ability to support your own body weight

RESTRICTIONS
• Those with lower-back problems should avoid this exercise.

③ Raise one leg and hold for 30 seconds, then switch legs.

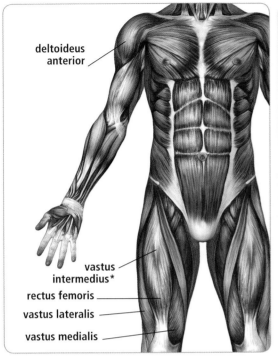
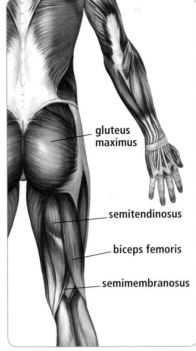

deltoideus
anterior

vastus
intermedius*

rectus femoris

vastus lateralis

vastus medialis

gluteus
maximus

semitendinosus

biceps femoris

semimembranosus

TARGET AREAS

Primary emphasis is on
the glutes, hamstrings,
quadriceps, deltoids,
biceps, triceps, abdominals,
and erectors.

ANNOTATION KEY
Bold text indicates target muscles
Gray text indicates other working muscles
* indicates deep muscles

transversus abdominis*

tensor fasciae
latae

rectus
abdominis

adductor longus

obliquus
externus

adductor magnus

rectus
femoris

biceps
brachii

tibialis anterior

obliquus
internus*

peroneus

biceps femoris

gluteus maximus

gluteus
medius*

triceps
brachii

PLANK ROLL-DOWN

1 From a standing position, bend at the waist, keeping your legs straight as you touch the floor with your hands.

2 Walk your hands away from your feet until you have reached a Plank position on your toes.

DO IT RIGHT
Keep your body straight when in the Plank position.

AVOID
Dipping too low, since this can strain the lower back.

LEVEL
• Intermediate

TIME
• 1-minute completion time

BENEFITS
• Helps strengthen and stabilize the upper body

RESTRICTIONS
• Those with limited wrist mobility or shoulder pain should avoid this exercise.

3 Once you are in the Plank position, keep your arms straight as you dip your shoulders.

4 Hold for 10 seconds, then walk your hands back to your feet and return to an upright position. Repeat six times.

MODIFICATION

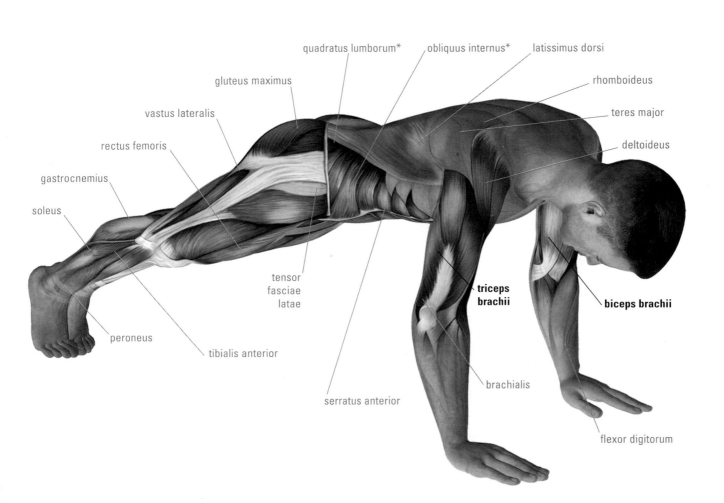

This exercise can be made easier by keeping your forearms planted on the floor, instead of your hands.

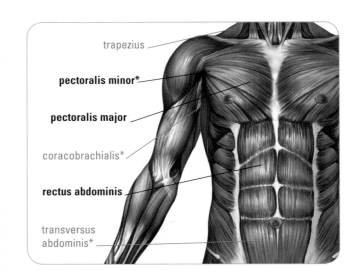

- trapezius
- **pectoralis minor***
- **pectoralis major**
- coracobrachialis*
- **rectus abdominis**
- transversus abdominis*

ANNOTATION KEY

Bold text indicates target muscles
Gray text indicates other working muscles
* indicates deep muscles

TARGET AREAS

Primary emphasis is on the pectorals, biceps, triceps, and rectus abdominis.

- quadratus lumborum*
- obliquus internus*
- latissimus dorsi
- gluteus maximus
- rhomboideus
- vastus lateralis
- teres major
- rectus femoris
- deltoideus
- gastrocnemius
- soleus
- tensor fasciae latae
- **triceps brachii**
- **biceps brachii**
- peroneus
- tibialis anterior
- brachialis
- serratus anterior
- flexor digitorum

WALL SITS

① Stand against a wall. Then, while leaning against it, take a step forward with both feet, and push your lower back into the wall.

DO IT RIGHT
Keep your lower back pressed against the wall at all times.

AVOID
Letting your knees extend past your feet.

② Slide down the wall, as if performing a squat, until your thighs are roughly parallel to the ground. Raise your arms so they are also parallel to the ground, and hold the pose for 60 seconds. Complete five repetitions.

LEVEL
• Intermediate

TIME
• 5-minute completion time

BENEFITS
• Strengthens the lower body

RESTRICTIONS
• Those with knee pain should avoid this exercise.

TARGET AREAS

Primary emphasis is on the abdominals, quads, glutes, and hamstrings

ANNOTATION KEY

Bold text indicates target muscles

Gray text indicates other working muscles

* indicates deep muscles

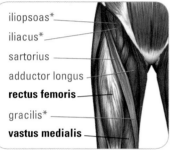

gluteus medius*

adductor magnus

biceps femoris

semitendinosus

semimembranosus

iliopsoas*

iliacus*

sartorius

adductor longus

rectus femoris

gracilis*

vastus medialis

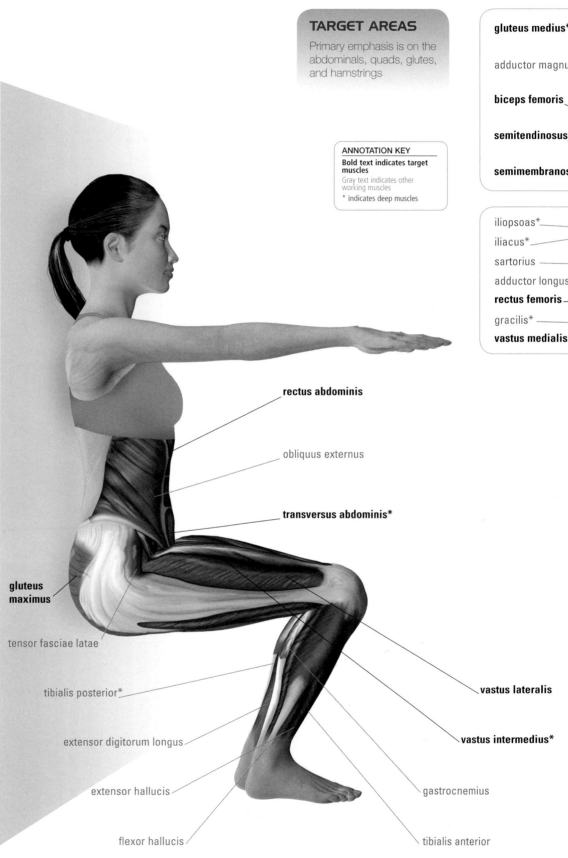

rectus abdominis

obliquus externus

transversus abdominis*

gluteus maximus

tensor fasciae latae

tibialis posterior*

extensor digitorum longus

extensor hallucis

flexor hallucis

vastus lateralis

vastus intermedius*

gastrocnemius

tibialis anterior

QUADRUPED

DO IT RIGHT	AVOID
Maintain a flat back throughout the exercise.	Using any jerky movements.

1 Position yourself on all fours, with your hands and feet shoulder-width apart.

LEVEL
• Beginner

TIME
• 3-minute completion time

BENEFITS
• Helps develop the core

RESTRICTIONS
• Those with lower-back pain should avoid this exercise.

2 Fully extend your left leg behind you while straightening the right arm out in front. Hold for 30 seconds, then return to the starting position on all fours. Repeat the exercise, then switch to the other leg and arm for two more repetitions.

MODIFICATION

To make this exercise more difficult, first assume the modified Plank position (see page 41), then continue with step 2 (opposite).

TARGET AREAS

Primary emphasis is on the shoulders, upper back, and core.

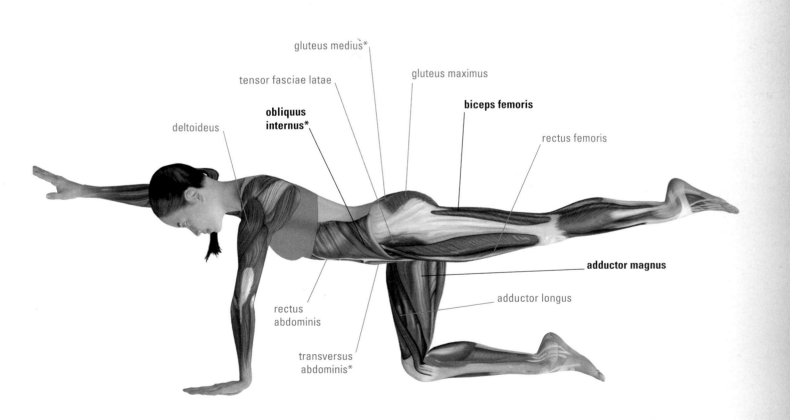

gluteus medius*

tensor fasciae latae

gluteus maximus

obliquus internus*

biceps femoris

deltoideus

rectus femoris

adductor magnus

rectus abdominis

adductor longus

transversus abdominis*

TRANSVERSE ABS

❶ Begin positioned on your toes, with your elbows and forearms on top of a Swiss ball.

❷ Keep your body in a straight line, and hold for 30–60 seconds.

LEVEL
• Advanced

TIME
• 60-second completion time

BENEFITS
• Will improve core strength and stability

RESTRICTIONS
• Those with lower-back problems should avoid this exercise.

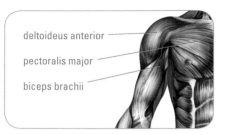

deltoideus anterior

pectoralis major

biceps brachii

ANNOTATION KEY
Bold text indicates target muscles
Gray text indicates other working muscles
* indicates deep muscles

TARGET AREAS
Primary emphasis is on the entire core.

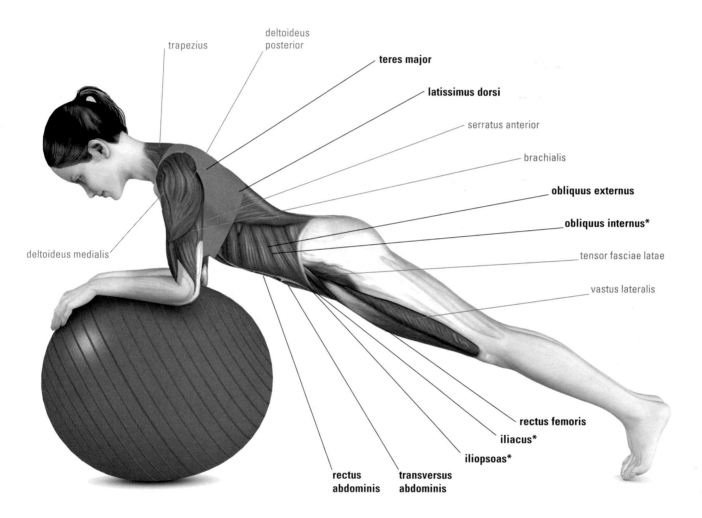

trapezius

deltoideus posterior

teres major

latissimus dorsi

serratus anterior

brachialis

obliquus externus

obliquus internus*

tensor fasciae latae

vastus lateralis

deltoideus medialis

rectus femoris

iliacus*

iliopsoas*

rectus abdominis

transversus abdominis

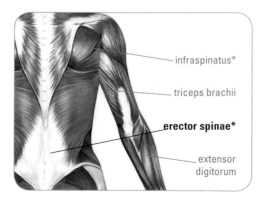

infraspinatus*

triceps brachii

erector spinae*

extensor digitorum

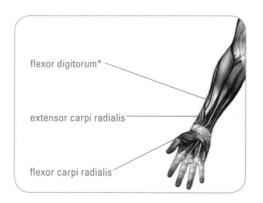

flexor digitorum*

extensor carpi radialis

flexor carpi radialis

SWISS BALL SIT TO BRIDGE

① Sit upright on the Swiss ball, with your feet flat on the floor and your hands resting on your knees or thighs.

DO IT RIGHT	AVOID
Lean back slowly and with control, and exhale while contracting.	Allowing the ball to shift laterally.

② Extend your arms in front of you, and slowly step forward while leaning back on the ball, allowing it to roll up your spine.

LEVEL
• Intermediate

TIME
• 30-second completion time

BENEFITS
• Increases spinal extension, and stretches upper back and abs

RESTRICTIONS
• Those with lower-back problems should avoid this exercise.

③ Walk your feet forward, so that the ball continues to roll up your spine, simultaneously extending your arms back and over your head.

④ Roll back until your hands are on the floor, with your arms slightly bent, and the back of your head rests against the ball. Hold this bridged position for 5 seconds, ending in an exhale.

⑤ To release the stretch, lift your head from the ball, and then slowly walk backward to the starting position.

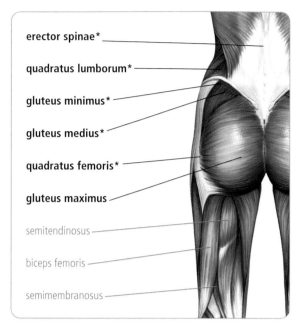

erector spinae*

quadratus lumborum*

gluteus minimus*

gluteus medius*

quadratus femoris*

gluteus maximus

semitendinosus

biceps femoris

semimembranosus

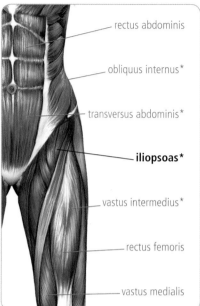

rectus abdominis

obliquus internus*

transversus abdominis*

iliopsoas*

vastus intermedius*

rectus femoris

vastus medialis

TARGET AREAS

Primary emphasis is on the entire core.

ANNOTATION KEY

Bold text indicates target muscles

Gray text indicates other working muscles

* indicates deep muscles

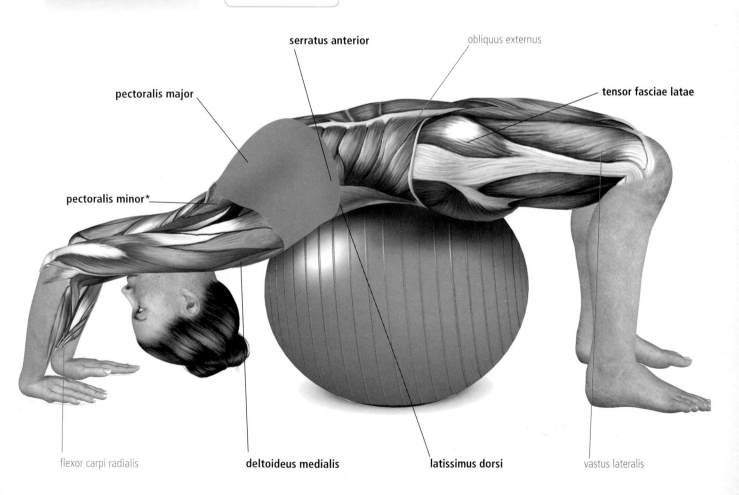

serratus anterior

obliquus externus

pectoralis major

tensor fasciae latae

pectoralis minor*

flexor carpi radialis

deltoideus medialis

latissimus dorsi

vastus lateralis

BRIDGE

① Begin on your back with your legs bent, your feet flat on the ground, and your arms extended on the floor, parallel to your body.

DO IT RIGHT	AVOID
Push through your heels, not your toes.	Overextending your abdominals past your thighs in the finished position.

LEVEL
- Beginner

TIME
- 90-second completion time

BENEFITS
- Increases strength in glutes and hamstrings

RESTRICTIONS
- Those with lower-back problems should avoid this exercise.

② Push through your heels while raising your pelvis until your torso is aligned with your thighs. Hold for 30 seconds, then lower yourself back down. Perform three repetitions.

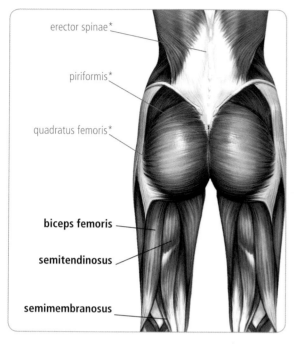

erector spinae*
piriformis*
quadratus femoris*
biceps femoris
semitendinosus
semimembranosus

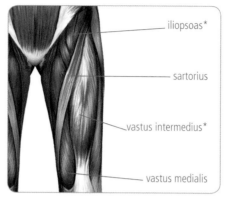

iliopsoas*
sartorius
vastus intermedius*
vastus medialis

TARGET AREAS

Primary emphasis is on the glute and hamstring muscles.

ANNOTATION KEY
Bold text indicates target muscles
Gray text indicates other working muscles
* indicates deep muscles

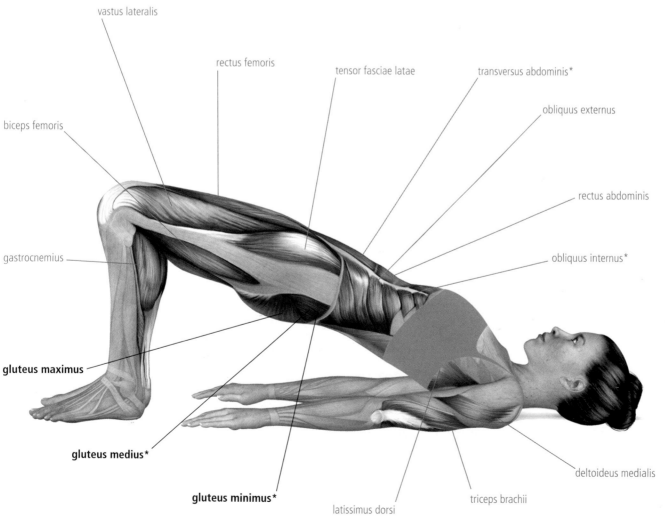

vastus lateralis
rectus femoris
tensor fasciae latae
transversus abdominis*
obliquus externus
rectus abdominis
biceps femoris
obliquus internus*
gastrocnemius
gluteus maximus
deltoideus medialis
gluteus medius*
triceps brachii
gluteus minimus*
latissimus dorsi

SINGLE-LEG GLUTE BRIDGE

STATIC

DO IT RIGHT	AVOID
Keep your back pressed to the ground.	Excessively pulling on or straining the knee.

1 Begin on your back with your legs bent, your feet flat on the ground, and your arms extended along your sides.

LEVEL
• Intermediate

TIME
• 2-minute completion time

BENEFITS
• Increases strength in glutes and hamstrings

RESTRICTIONS
• Those with lower-back problems should avoid this exercise.

2 Raise your left foot off the floor, keeping your knee bent at a 90-degree angle, until your thigh is perpendicular to your torso.

3 Push through your right heel while raising your pelvis until your torso is aligned with the planted thigh. Hold for 30 seconds, repeat, and then switch legs.

58

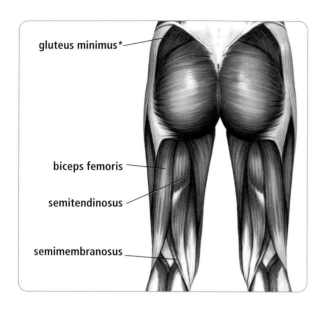

- gluteus minimus*
- biceps femoris
- semitendinosus
- semimembranosus

TARGET AREAS

Primary emphasis is on the glute and hamstring muscles.

ANNOTATION KEY
Bold text indicates target muscles
Gray text indicates other working muscles
* indicates deep muscles

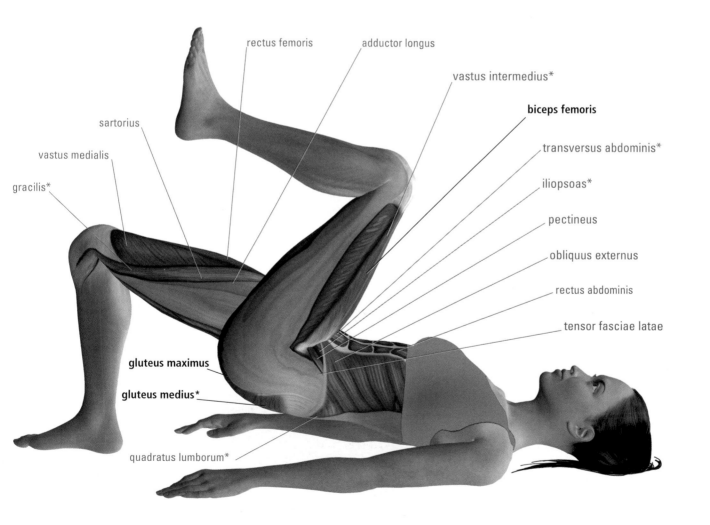

- rectus femoris
- adductor longus
- vastus intermedius*
- **biceps femoris**
- sartorius
- transversus abdominis*
- vastus medialis
- iliopsoas*
- gracilis*
- pectineus
- obliquus externus
- rectus abdominis
- tensor fasciae latae
- **gluteus maximus**
- **gluteus medius***
- quadratus lumborum*

BOTTOM PUSH-UP HOLD

DO IT RIGHT	AVOID
Keep your chest and abdominal muscles active.	Bridging too high, since this can take stress off the working muscles.

1 Start facedown on your tiptoes and palms. Your hands should be parallel to one another and just beyond shoulder-width apart, as if you were about to perform a push-up.

2 Raise your knees and chest, and extend your legs. Remain suspended in this bottom push-up position for 30 seconds (building up to 120 seconds).

LEVEL
• Advanced

TIME
• 2-minute completion time

BENEFITS
• Increases the ability to support your own body weight

RESTRICTIONS
• Those with lower-back problems should avoid this exercise.

MODIFICATION
• The Bottom Push-up Hold can be made easier by keeping your knees on the ground.

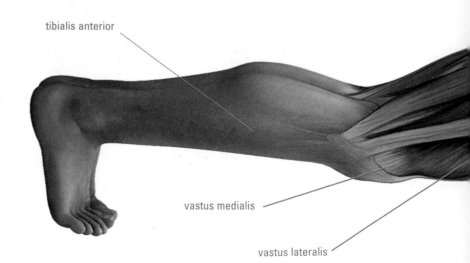

tibialis anterior

vastus medialis

vastus lateralis

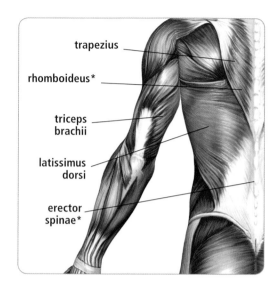

trapezius

rhomboideus*

triceps brachii

latissimus dorsi

erector spinae*

TARGET AREAS

Primary emphasis is on the pectorals, anterior deltoids, upper back, triceps, and the core.

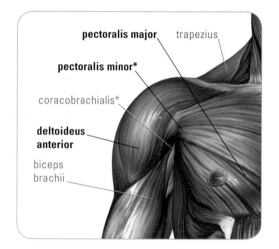

pectoralis major   trapezius

pectoralis minor*

coracobrachialis*

deltoideus anterior

biceps brachii

ANNOTATION KEY

Bold text indicates target muscles

Gray text indicates other working muscles

* indicates deep muscles

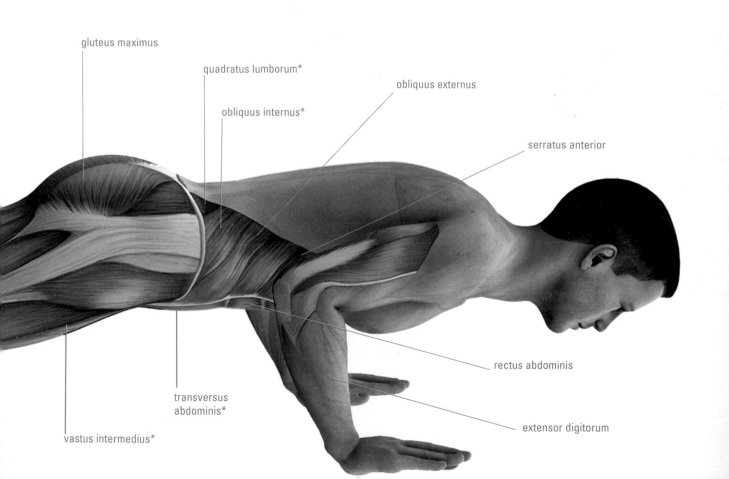

gluteus maximus

quadratus lumborum*

obliquus internus*

obliquus externus

serratus anterior

rectus abdominis

transversus abdominis*

extensor digitorum

vastus intermedius*

SINGLE-LEG BALANCE

STATIC

① Stand with your hands on your hips, and raise your right leg, bent at the knee, directly in front of you at a 90-degree angle. Hold for 15 seconds.

DO IT RIGHT	AVOID
Maintain an erect posture throughout the exercise.	Removing your hands from your hips.

② Press your right leg down and forward, though not touching the floor, and hold for 15 seconds.

③ Finally, press your right leg out to the side, again without touching the floor, and hold for 15 seconds. Complete the entire sequence three times, then switch legs.

LEVEL
• Intermediate

TIME
• 5-minute completion time

BENEFITS
• Strengthens the legs and core, and increases your stability

RESTRICTIONS
• Those with knee problems should avoid this exercise.

MODIFICATION
• This exercise can be made more difficult by gently tapping your heel to the floor between steps.

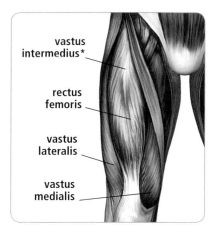

vastus intermedius*

rectus femoris

vastus lateralis

vastus medialis

TARGET AREAS

Primary emphasis is on the entire core and the quadriceps and hamstrings.

ANNOTATION KEY

Bold text indicates target muscles
Gray text indicates other working muscles
* indicates deep muscles

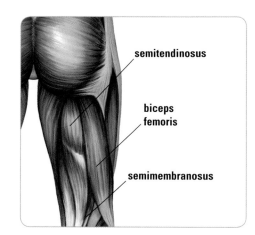

semitendinosus

biceps femoris

semimembranosus

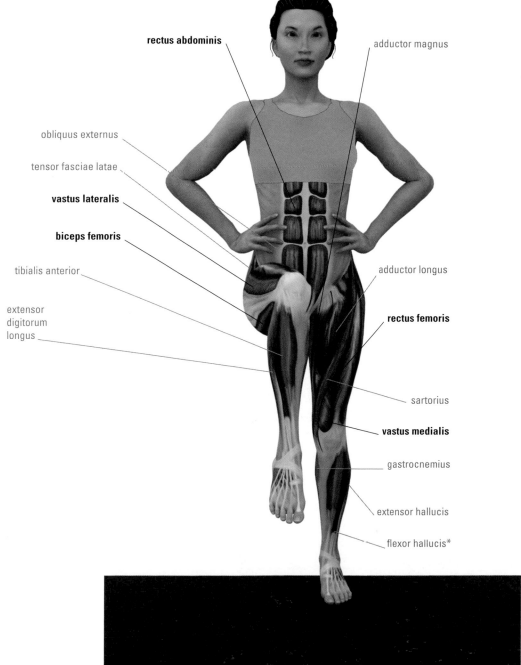

rectus abdominis

adductor magnus

obliquus externus

tensor fasciae latae

vastus lateralis

biceps femoris

tibialis anterior

extensor digitorum longus

adductor longus

rectus femoris

sartorius

vastus medialis

gastrocnemius

extensor hallucis

flexor hallucis*

HIGH LUNGE

1 Start in a standing position, and move your right foot forward while placing your hands on the floor, one on each side of the right foot.

DO IT RIGHT	AVOID
Maintain a flat back throughout the exercise to help keep your spine lengthened.	Letting your back knee touch the ground.

2 Take a giant step backward with the left leg so that it forms a straight line with your body. Keep the ball of the back foot in contact with the floor. Press through your right heel, contracting your thigh, and hold for 30 seconds (working up to 1 minute).

LEVEL
• Intermediate

TIME
• 1-minute completion time

BENEFITS
• Helps strengthen the abdominals and legs

RESTRICTIONS
• Those with a hip injury should avoid this exercise.

3 Bring your left leg back in line with your right, then repeat the exercise with your right leg stepping backward.

gastrocnemius

64

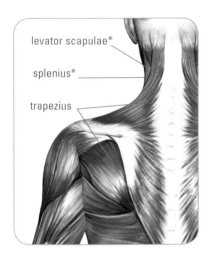

levator scapulae*

splenius*

trapezius

TARGET AREAS

Primary emphasis is on the entire core, glutes, quadriceps, hamstrings, and calves.

ANNOTATION KEY
Bold text indicates target muscles
Gray text indicates other working muscles
* indicates deep muscles

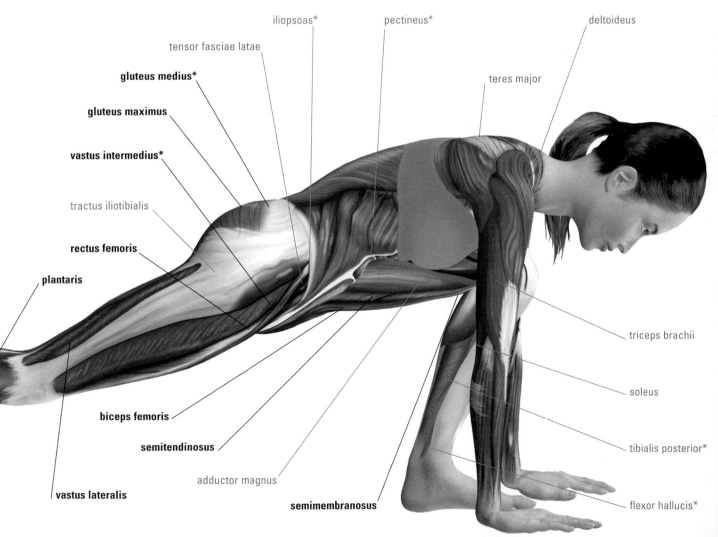

iliopsoas*

pectineus*

deltoideus

tensor fasciae latae

gluteus medius*

teres major

gluteus maximus

vastus intermedius*

tractus iliotibialis

rectus femoris

plantaris

triceps brachii

soleus

biceps femoris

semitendinosus

tibialis posterior*

adductor magnus

vastus lateralis

semimembranosus

flexor hallucis*

65

INVERTED HAMSTRING

1 Begin in a standing position, feet shoulder-width apart, with your legs slightly bent and your arms above your head.

AVOID
Letting your foot touch the ground.

2 Bend forward at the waist while simultaneously spreading your arms out to your sides for balance and lifting your left leg behind you, until your torso and leg are roughly parallel to the ground. Hold for 15 seconds, and repeat.

LEVEL
• Advanced

TIME
• 1-minute completion time

BENEFITS
• Helps stabilize the body as a whole

RESTRICTIONS
• Those with lower-back problems should avoid this exercise.

MODIFICATION
• This exercise can be made easier by holding a balance pole out in front of you.

3 Return to a standing position, switch legs, and repeat step 2.

66

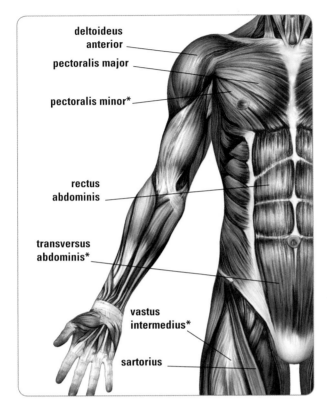

deltoideus
anterior

pectoralis major

pectoralis minor*

rectus
abdominis

transversus
abdominis*

vastus
intermedius*

sartorius

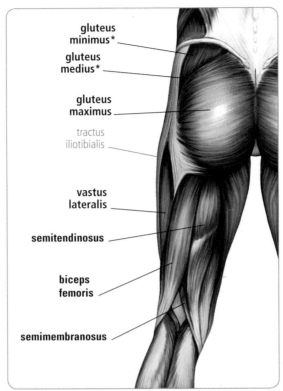

gluteus
minimus*

gluteus
medius*

gluteus
maximus

tractus
iliotibialis

vastus
lateralis

semitendinosus

biceps
femoris

semimembranosus

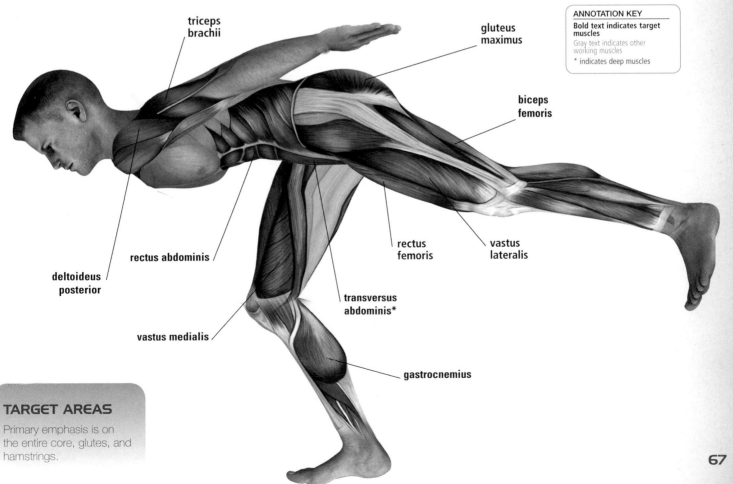

triceps
brachii

gluteus
maximus

biceps
femoris

rectus abdominis

deltoideus
posterior

rectus
femoris

vastus
lateralis

transversus
abdominis*

vastus medialis

gastrocnemius

ANNOTATION KEY
Bold text indicates target muscles
Gray text indicates other working muscles
* indicates deep muscles

TARGET AREAS

Primary emphasis is on the entire core, glutes, and hamstrings.

67

STATIC SUMO SQUAT

❶ Stand upright with your feet more than shoulder-width apart.

DO IT RIGHT
Maintain a neutral spine position.

AVOID
Letting your knees extend past your feet.

❷ Drop down into a deep squat with your hands resting on your inner thighs. Hold for 30 seconds, and repeat.

LEVEL
• Beginner

TIME
• 1-minute completion time

BENEFITS
• Helps keep the legs flexible and strong

RESTRICTIONS
• Those with problems in the lower back or knees should avoid this exercise.

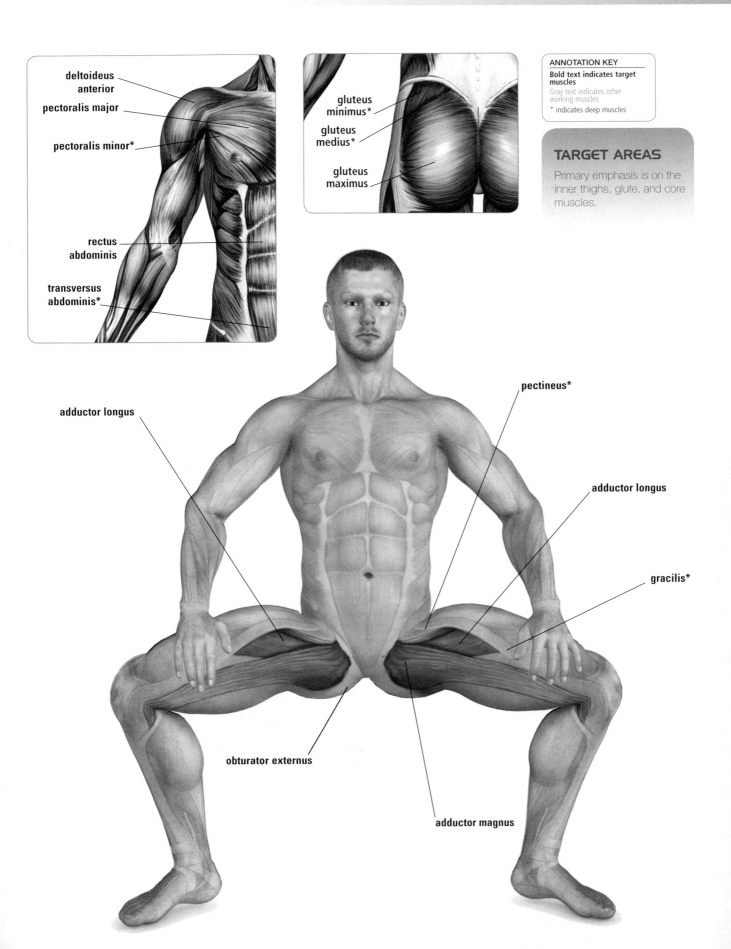

deltoideus anterior

pectoralis major

pectoralis minor*

rectus abdominis

transversus abdominis*

gluteus minimus*

gluteus medius*

gluteus maximus

ANNOTATION KEY

Bold text indicates target muscles
Gray text indicates other working muscles
* indicates deep muscles

TARGET AREAS

Primary emphasis is on the inner thighs, glute, and core muscles.

pectineus*

adductor longus

adductor longus

gracilis*

obturator externus

adductor magnus

LATERAL ROLL

<table>
<tr><td>DO IT RIGHT</td><td>AVOID</td></tr>
<tr><td>Keep your back braced.</td><td>Dropping your hips.</td></tr>
</table>

① Lie on a Swiss ball with your upper back firmly supported. Keep your feet flat on the floor, shoulder-width apart, your hips raised, and your arms stretched out to your sides.

LEVEL
• Intermediate

TIME
• 4-minute completion time

BENEFITS
• Helps stabilize the trunk

RESTRICTIONS
• Those who suffer from chronic lower-back pain should approach this exercise with caution.

② Roll the ball to the side in baby steps, and then walk back in the opposite direction. Complete three sets of 10 steps per side.

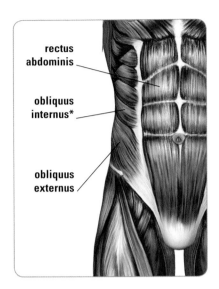

rectus
abdominis

obliquus
internus*

obliquus
externus

TARGET AREAS

Primary emphasis is on
the rectus abdominis and
obliques.

ANNOTATION KEY
**Bold text indicates target
muscles**
Gray text indicates other
working muscles
* indicates deep muscles

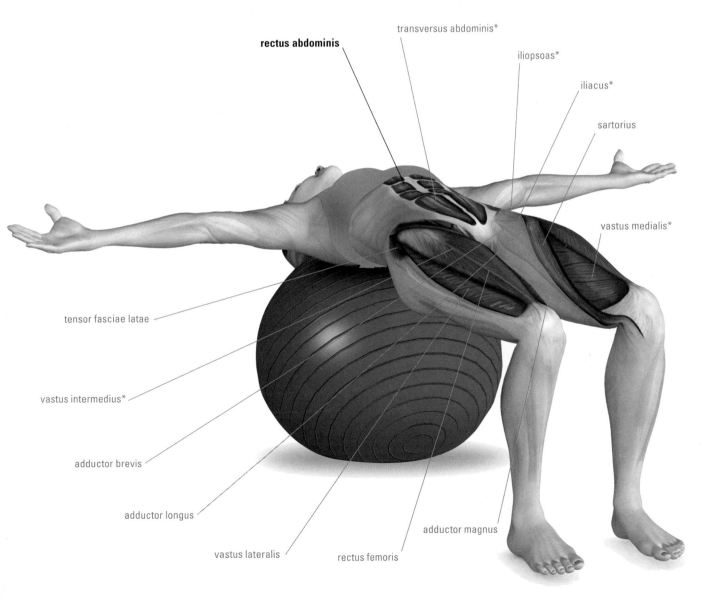

rectus abdominis

transversus abdominis*

iliopsoas*

iliacus*

sartorius

vastus medialis*

tensor fasciae latae

vastus intermedius*

adductor brevis

adductor longus

vastus lateralis

rectus femoris

adductor magnus

SWISS BALL HYPEREXTENSION

❶ Lie facedown on top of the Swiss ball, with your abdominals covering most of the ball and your hands on the floor.

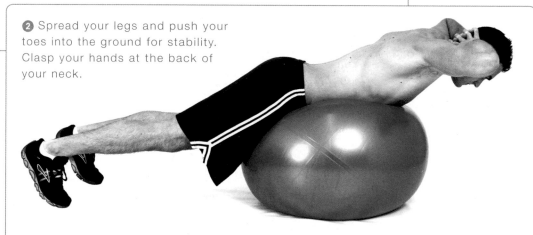

❷ Spread your legs and push your toes into the ground for stability. Clasp your hands at the back of your neck.

LEVEL
• Intermediate

TIME
• 30-second completion time

BENEFITS
• Strengthens the lower-back and glute areas

RESTRICTIONS
• Those with post-surgery lower-back problems should avoid this exercise.

MODIFICATION
• Beginners can perform this exercise with their feet braced against a wall for extra stability.

❸ Raise your torso so that it is in a straight line with the lower half of your body.

❹ Squeeze the glute muscles, and hold for 10 seconds. Perform three repetitions.

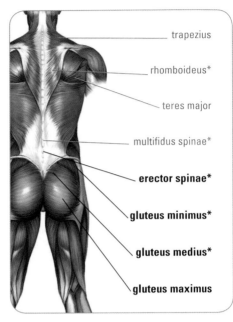

trapezius

rhomboideus*

teres major

multifidus spinae*

erector spinae*

gluteus minimus*

gluteus medius*

gluteus maximus

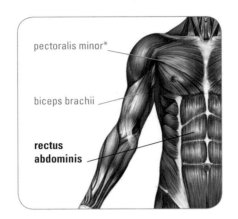

pectoralis minor*

biceps brachii

rectus abdominis

ANNOTATION KEY

Bold text indicates target muscles

Gray text indicates other working muscles

* indicates deep muscles

TARGET AREAS

Primary emphasis is on the glutes, erector spinae, and rectus abdominis.

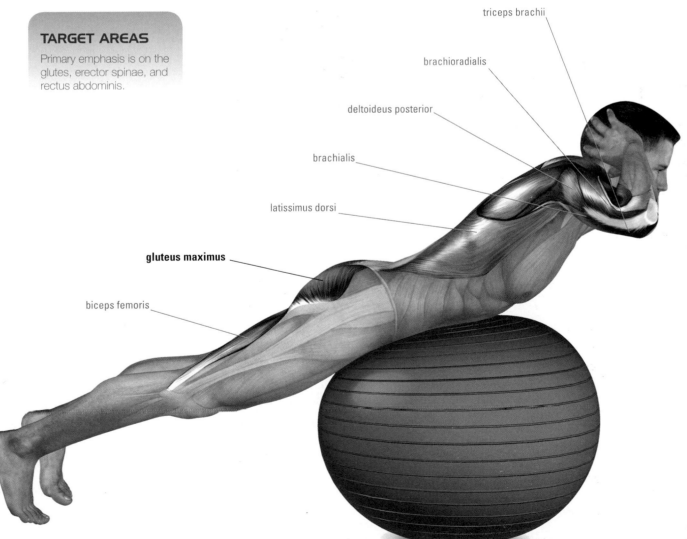

triceps brachii

brachioradialis

deltoideus posterior

brachialis

latissimus dorsi

gluteus maximus

biceps femoris

ROTATED BACK EXTENSION

DO IT RIGHT

Keep your hips square throughout the exercise.

AVOID

Overcontracting or hyperextending your back at the top of the movement.

1 Lie facedown on top of the Swiss ball, with your abdominals covering most of the ball. Spread your legs and push your toes into the ground for stability. Clasp your hands behind your head, with your fingers interlocked.

2 Raise your torso so that it is in line with the lower half of your body. At the same time, rotate it to the right.

LEVEL
• Advanced

TIME
• 1-minute completion time

BENEFITS
• Strengthens the lower back and oblique muscles

RESTRICTIONS
• Those with neck and/or lower-back pain should avoid this exercise.

3 Squeeze the glute muscles at the top, and hold for 10 seconds. Complete three repetitions per side.

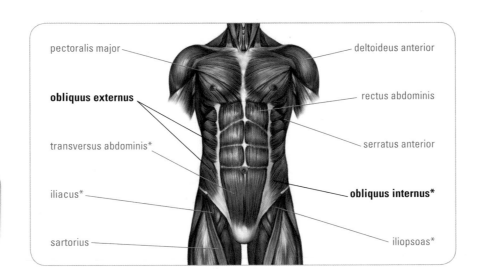

TARGET AREAS

Primary emphasis is on the erector spinae and obliques.

pectoralis major

deltoideus anterior

obliquus externus

rectus abdominis

transversus abdominis*

serratus anterior

iliacus*

obliquus internus*

sartorius

iliopsoas*

deltoideus medialis

extensor digitorum

deltoideus posterior

infraspinatus*

subscapularis*

rhomboideus*

erector spinae*

latissimus dorsi

tensor fasciae latae

rectus femoris

tibialis anterior

triceps brachii

brachialis

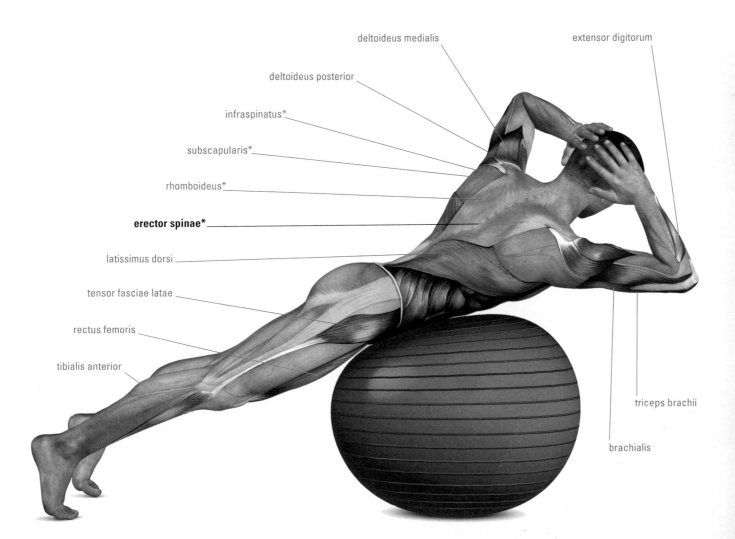

SIDE-LYING HIP ABDUCTION

STATIC

1 Lie on your left side with your legs extended and your feet stacked one on top of the another. Rest your right arm along your right hip, and use your left arm to support your head.

LEVEL
• Beginner

TIME
• 2-minute completion time

BENEFITS
• Improves glute and hip strength

RESTRICTIONS
• Those with lower-back problems should avoid this exercise.

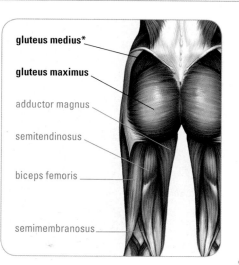

gluteus medius*

gluteus maximus

adductor magnus

semitendinosus

biceps femoris

semimembranosus

DO IT RIGHT
Keep your body in a straight line.

AVOID
Raising your leg too high.

ANNOTATION KEY
Bold text indicates target muscles
Gray text indicates other working muscles
* indicates deep muscles

vastus lateralis

vastus intermedius*

vastus medialis

TARGET AREAS

Primary emphasis is on the glutes and hips, with secondary assistance from the core.

2 Raise your right leg until you feel your core kick in. Hold for 30 seconds, lower, and repeat, then switch sides.

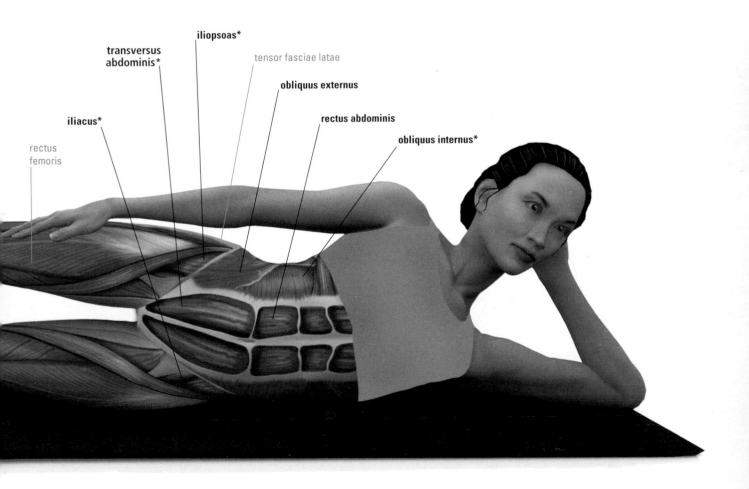

rectus femoris

iliacus*

transversus abdominis*

iliopsoas*

tensor fasciae latae

obliquus externus

rectus abdominis

obliquus internus*

TINY STEPS

STATIC

DO IT RIGHT	AVOID
Keep your abdominals pulled in during the exercise.	Moving your hips.

❶ Begin by lying on your back with your knees bent and your feet on tiptoes on the floor. Place your hands on your hipbones, and raise your left knee toward your chest while keeping your abdominals in close to your spine.

❷ As you lower your left leg to the floor, keep your abdominals tightened and hold for 10 seconds.

LEVEL
• Beginner

TIME
• 2-minute completion time

BENEFITS
• Increases lower-abdominal stability and helps protect the lower back

RESTRICTIONS
• Those with lower-back problems should avoid this exercise.

❸ Switch legs, and repeat six times per side.

78

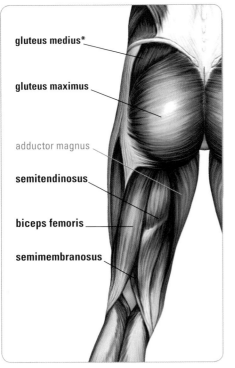

- gluteus medius*
- gluteus maximus
- adductor magnus
- semitendinosus
- biceps femoris
- semimembranosus

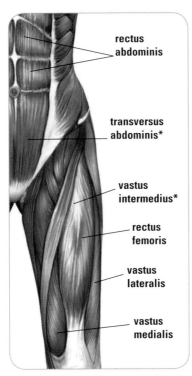

- rectus abdominis
- transversus abdominis*
- vastus intermedius*
- rectus femoris
- vastus lateralis
- vastus medialis

TARGET AREAS

Primary emphasis is on the lower abdominals, glutes, quads, and hamstrings.

ANNOTATION KEY

Bold text indicates target muscles
Gray text indicates other working muscles
* indicates deep muscles

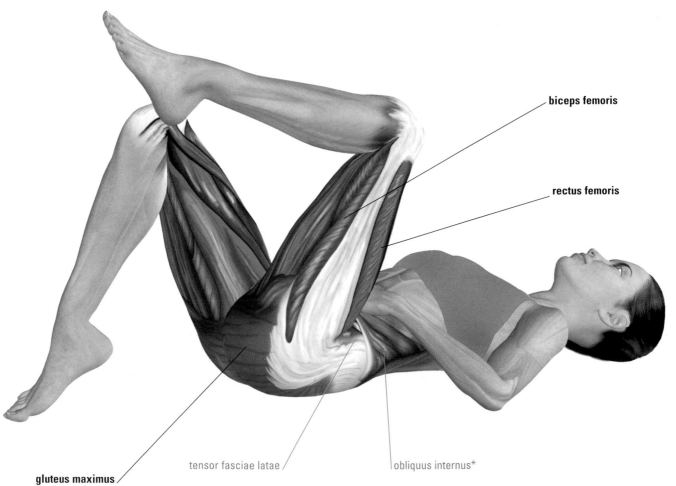

- biceps femoris
- rectus femoris
- gluteus maximus
- tensor fasciae latae
- obliquus internus*

DOUBLE-LEG AB PRESS

STATIC

DO IT RIGHT

Keep your feet flexed and your knees pressed together.

AVOID

Holding your breath during this exercise.

❶ Lie on your back with both legs raised and bent to a 90-degree angle and your hands on your knees.

LEVEL
• Beginner

TIME
• 5-minute completion time

BENEFITS
• Strengthens the core, hip flexors, and triceps

RESTRICTIONS
• Those with lower-back problems should avoid this exercise.

MODIFICATION
• Beginners can plant their feet against a hard surface for extra support.

❷ While keeping your feet flexed, perform an ab crunch, lifting your shoulders and head from the floor, and press your hands into your knees. Simultaneously, push against your hands with your knees to create resistance. Hold for 60 seconds. Perform five repetitions.

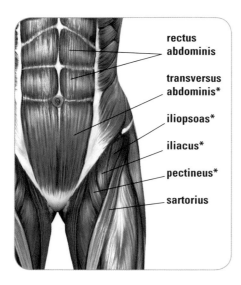

- **rectus abdominis**
- **transversus abdominis***
- **iliopsoas***
- **iliacus***
- **pectineus***
- **sartorius**

TARGET AREAS

Primary emphasis is on the core, hip flexors, and lower back.

ANNOTATION KEY
Bold text indicates target muscles
Gray text indicates other working muscles
* indicates deep muscles

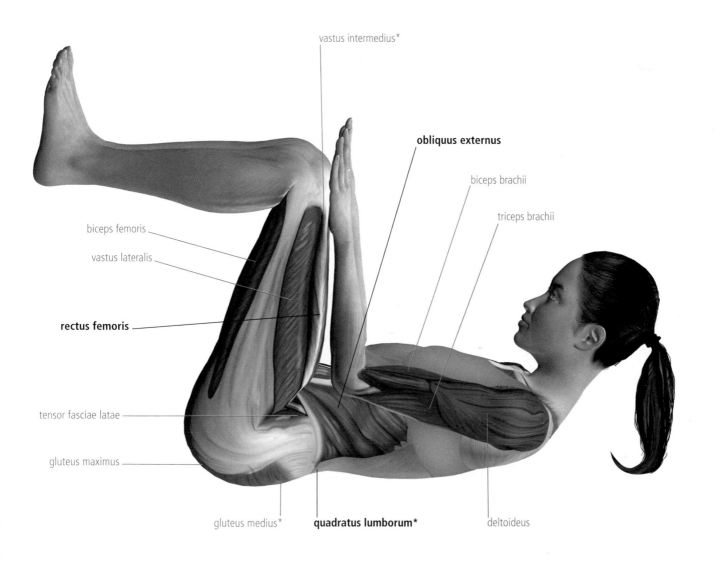

vastus intermedius*

obliquus externus

biceps brachii

triceps brachii

biceps femoris

vastus lateralis

rectus femoris

tensor fasciae latae

gluteus maximus

gluteus medius*

quadratus lumborum*

deltoideus

DYNAMIC EXERCISES

Dynamic exercises keep both joints and muscles moving, and they work

the abdominal and stabilizer muscles. Motions also occur through the

lumbar spine. Examples of dynamic exercise include swimming, walking,

cross-country skiing, bicycling, weight training, and even housework.

These forms of exercise rely on varying degrees of motion and employ

the negative (or stretch) part of the movement, followed by the positive

(or contracted) part. In a squat, for example, the downward portion

of the movement is the negative, and the explosive upward motion

is the positive part.

SWISS BALL ROLL-OUT

① Kneel in front of a Swiss ball, and place your hands on it at about hip height.

LEVEL
• Intermediate

TIME
• 3-minute completion time

BENEFITS
• Helps you support and handle your own body weight

RESTRICTIONS
• Pregnant women and those with post-surgery lower-back problems should avoid this exercise.

② Start rolling the ball forward slowly, extending your body as you go.

3 Continue rolling forward until you are completely stretched out while maintaining a flat back and remaining anchored on your knees. Then, using your abdominals and lower back, roll back to the starting position. Perform three sets of 15 repetitions.

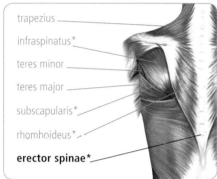

trapezius
infraspinatus*
teres minor
teres major
subscapularis*
rhomboideus*
erector spinae*

TARGET AREAS

Primary emphasis is on the abdominals and lower back.

ANNOTATION KEY
Bold text indicates target muscles
Gray text indicates other working muscles
* indicates deep muscles

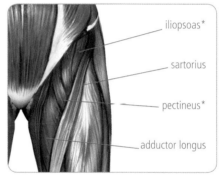

iliopsoas*
sartorius
pectineus*
adductor longus

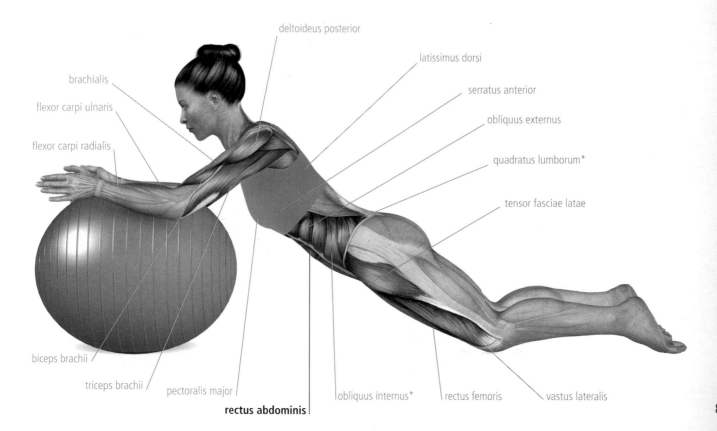

deltoideus posterior
latissimus dorsi
serratus anterior
obliquus externus
quadratus lumborum*
tensor fasciae latae

brachialis
flexor carpi ulnaris
flexor carpi radialis

biceps brachii
triceps brachii
pectoralis major
rectus abdominis
obliquus internus*
rectus femoris
vastus lateralis

SWISS BALL JACKKNIFE

1 Position yourself on all fours with your hands shoulder-width apart, then raise your left leg and place it on top of the Swiss ball.

DO IT RIGHT
Be sure to brace your core.

AVOID
Rounding your back.

LEVEL
• Advanced

TIME
• 2-minute completion time

BENEFITS
• Helps stabilize the torso

RESTRICTIONS
• Those who suffer from chronic lower-back pain should approach this exercise with caution.

2 Do the same with your right foot, too, so that you are in a push-up position with your shins resting on the Swiss ball.

3 Bend your knees, rolling the ball in toward your chest.

4 Bring your knees in toward your chest as far as you are able, then extend your legs back out to their starting position. Complete 20 repetitions.

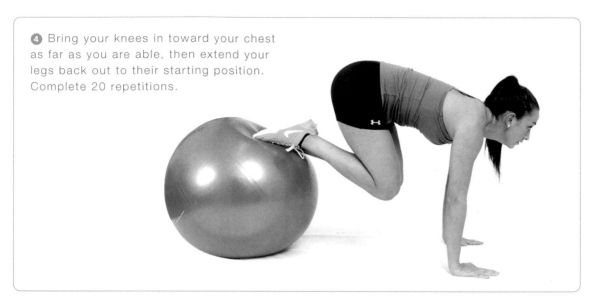

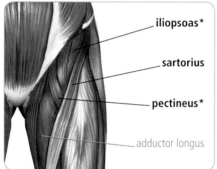

iliopsoas*

sartorius

pectineus*

adductor longus

TARGET AREAS

Primary emphasis is on the hip flexors, rectus abdominis, and erector spinae.

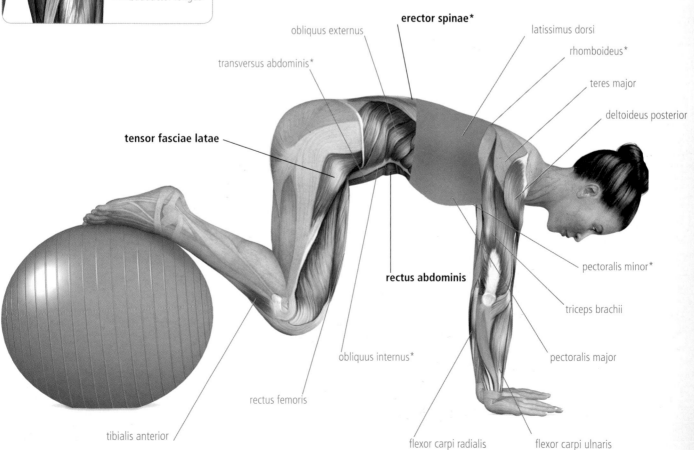

erector spinae*

obliquus externus

transversus abdominis*

latissimus dorsi

rhomboideus*

teres major

deltoideus posterior

tensor fasciae latae

pectoralis minor*

rectus abdominis

triceps brachii

obliquus internus*

pectoralis major

rectus femoris

tibialis anterior

flexor carpi radialis

flexor carpi ulnaris

SWISS BALL HIP CROSSOVER

1 Lie on your back, with your arms extended out to your sides. Place your legs on a Swiss ball, with your glutes close to it.

DO IT RIGHT
Keep the movement as smooth as possible.

AVOID
Swinging your legs excessively.

LEVEL
• Intermediate

TIME
• 3-minute completion time

BENEFITS
• Helps strengthen and tone the abs, and improves core stabilization

RESTRICTIONS
• Those with lower-back problems should avoid this exercise.

2 Brace your abdominal muscles, and lower your legs to one side, as close to the floor as you can without raising your shoulders off the floor.

3 Return to the starting position, and then switch to the other side. Work up to completing 20 repetitions in each direction.

TARGET AREAS

Primary emphasis is on the abdominals and lower back.

ANNOTATION KEY
Bold text indicates target muscles
Gray text indicates other working muscles
* indicates deep muscles

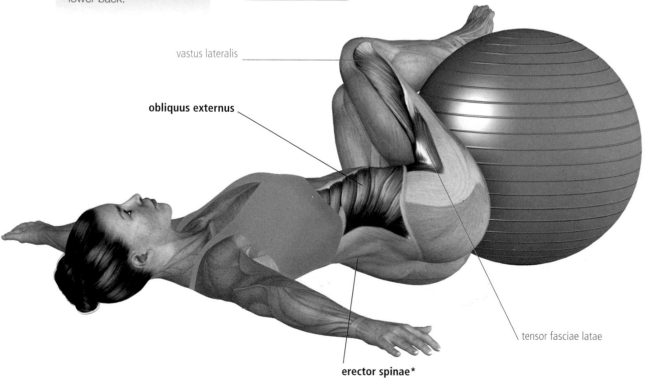

vastus lateralis

obliquus externus

tensor fasciae latae

erector spinae*

SWISS BALL WALK-AROUND

❶ Begin in a push-up position with a Swiss ball under your shins.

LEVEL
• Advanced

TIME
• 3-minute completion time

BENEFITS
• Helps keep the upper body stabilized and working cohesively

RESTRICTIONS
• Those with shoulder problems should avoid this exercise.

❷ "Walk" one hand at a time toward the right, turning the body with it until you have completed a semicircle. Then hand-walk to the left, returning to your starting position. Complete three semicircles in each direction.

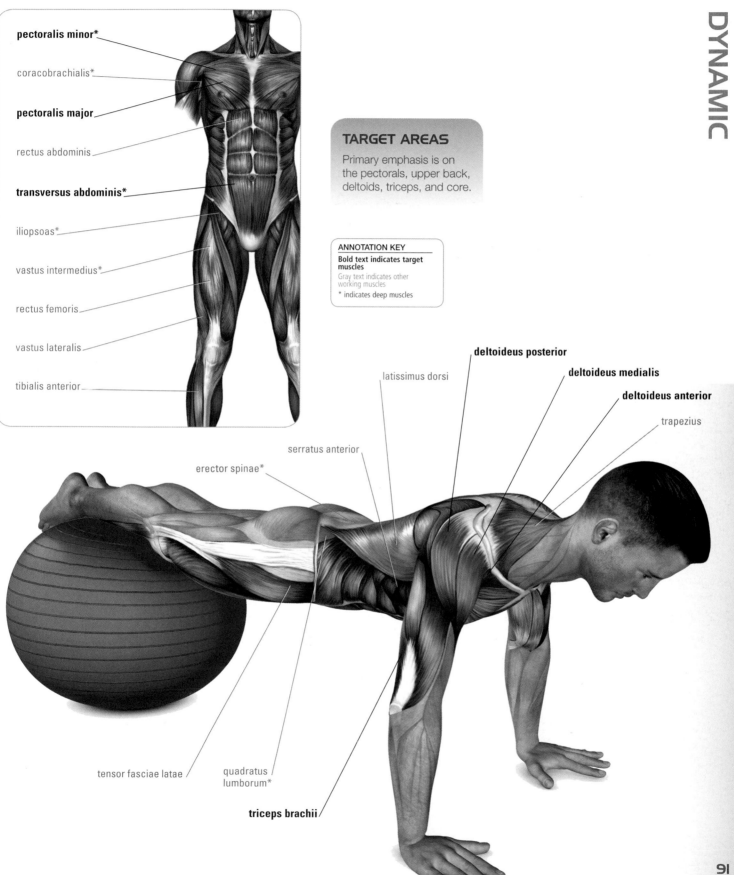

pectoralis minor*

coracobrachialis*

pectoralis major

rectus abdominis

transversus abdominis*

iliopsoas*

vastus intermedius*

rectus femoris

vastus lateralis

tibialis anterior

TARGET AREAS

Primary emphasis is on the pectorals, upper back, deltoids, triceps, and core.

ANNOTATION KEY

Bold text indicates target muscles

Gray text indicates other working muscles

* indicates deep muscles

deltoideus posterior

deltoideus medialis

latissimus dorsi

deltoideus anterior

trapezius

serratus anterior

erector spinae*

tensor fasciae latae

quadratus lumborum*

triceps brachii

PUSH-UP

❶ Begin facedown, with your hands planted on the floor, shoulder-width apart, and your arms fully extended. Lengthen your legs, and balance on your toes.

❷ Bend your arms until your chest is nearly touching the floor, then push back to full extension. Complete three sets of 10 repetitions.

LEVEL
• Intermediate

TIME
• 3-minute completion time

BENEFITS
• Helps keep the upper body strong and stabilized

RESTRICTIONS
• Those with shoulder and/ or lower-back problems should avoid this exercise.

DO IT RIGHT	AVOID
Keep your chest directly over your hands.	Arching your back.

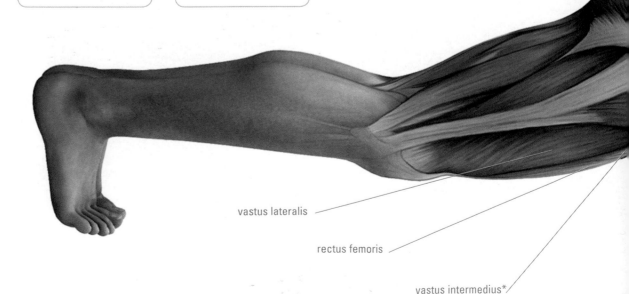

gluteus maximus

vastus lateralis

rectus femoris

vastus intermedius*

MODIFICATION

This exercise can be made easier by bracing your knees on the ground.

TARGET AREAS

Primary emphasis is on the pectorals, anterior deltoids, upper back, triceps, and core muscles.

ANNOTATION KEY
Bold text indicates target muscles
Gray text indicates other working muscles
* indicates deep muscles

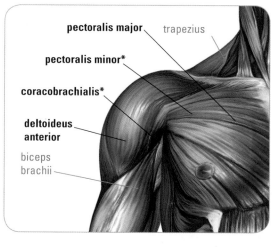

pectoralis major
trapezius
pectoralis minor*
coracobrachialis*
deltoideus anterior
biceps brachii

quadratus lumborum*
obliquus internus*
obliquus externus

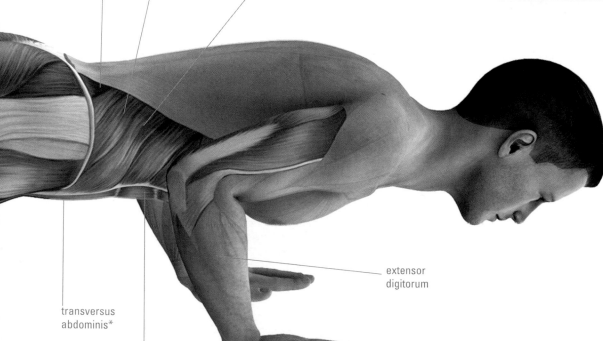

extensor digitorum

transversus abdominis*

rectus abdominis

PUSH-UP HAND WALKOVER

1 Begin in a top push-up position, with your hands palm-down, thumbs touching, on a stable, solid block.

DO IT RIGHT
Your chest should not drop below the height of the block.

AVOID
Excessive bouncing between repetitions.

LEVEL
• Advanced

TIME
• 2-minute completion time

BENEFITS
• Helps keep the upper body stabilized, active, and working cohesively

RESTRICTIONS
• Those with shoulder problems should avoid this exercise.

MODIFICATION
• This exercise can be modified by completing all reps on one side before switching hands.

2 Move your left hand from the block and place it on the floor as far to the left as is comfortable. Lower yourself to the ground as you would in a normal push-up.

94

3 As you push back up, move your left arm back to the block and repeat the movement with your right hand out to the side. Continue going back and forth until you have completed 30 push-ups in all.

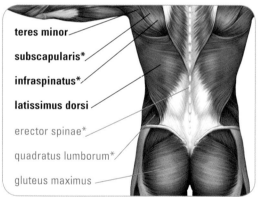

teres minor

subscapularis*

infraspinatus*

latissimus dorsi

erector spinae*

quadratus lumborum*

gluteus maximus

TARGET AREAS

Primary emphasis is on the pectorals, upper back, anterior shoulders, and triceps.

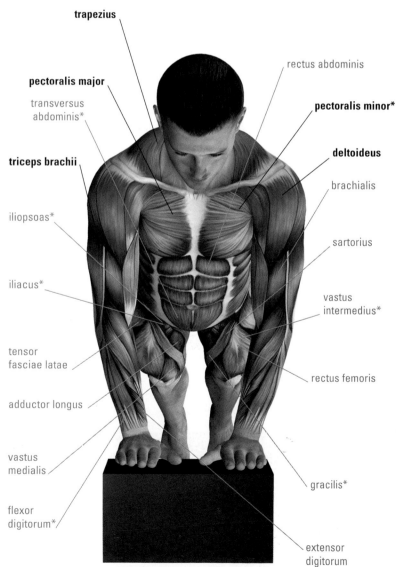

trapezius

pectoralis major

transversus abdominis*

triceps brachii

iliopsoas*

iliacus*

tensor fasciae latae

adductor longus

vastus medialis

flexor digitorum*

rectus abdominis

pectoralis minor*

deltoideus

brachialis

sartorius

vastus intermedius*

rectus femoris

gracilis*

extensor digitorum

CHAIR DIP

DYNAMIC

1 Stand close to, but facing away from, the front of a heavy chair (as if you were about to sit down). Bending at the knees, reach behind and place your hands on the leading edge of the chair. Scoot forward with each foot, until your knees are directly over your heels.

DO IT RIGHT
Keep your spine neutral throughout the exercise.

AVOID
Overusing your shoulders.

LEVEL
• Intermediate

TIME
• 3-minute completion time

BENEFITS
• Helps strengthen the shoulder girdle and stabilize the torso while the body is in motion

RESTRICTIONS
• Those with shoulder and/or wrist pain should avoid this exercise.

2 Gently lower yourself by bending the elbows until your upper arms are at a 90-degree angle to your forearms.

3 Push yourself back up until your arms are again fully extended. Complete three sets of 10 repetitions.

96

MODIFICATION

This exercise can be made more challenging by lifting one leg off the ground for the dip.

TARGET AREAS

Primary emphasis is on the triceps, deltoids, latissimus dorsi, rectus abdominis, and chest muscles.

ANNOTATION KEY

Bold text indicates target muscles

Gray text indicates other working muscles

* indicates deep muscles

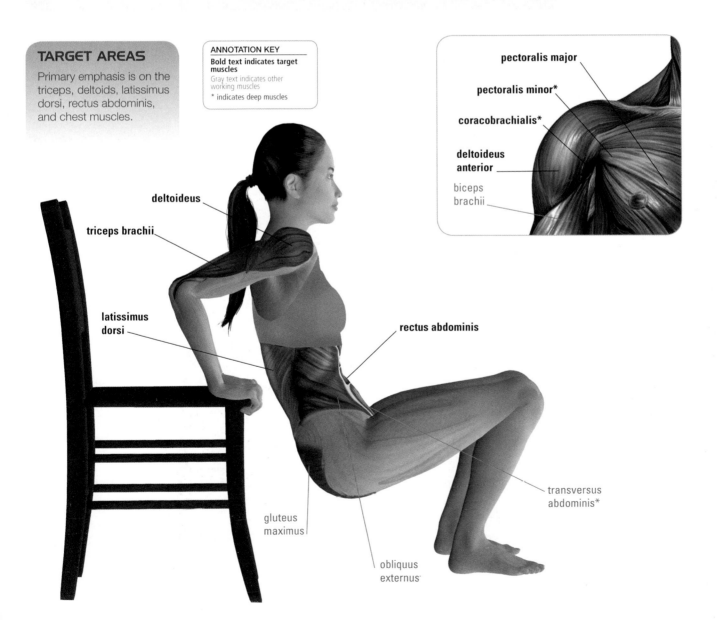

pectoralis major

pectoralis minor*

coracobrachialis*

deltoideus anterior

biceps brachii

deltoideus

triceps brachii

latissimus dorsi

rectus abdominis

transversus abdominis*

gluteus maximus

obliquus externus

TOWEL FLY

❶ Begin in a standard push-up position with a towel on the floor under your chest. Place your hands on the towel, a little more than shoulder-width apart.

LEVEL
• Advanced

TIME
• 2-minute completion time

BENEFITS
• Helps keep the upper body stabilized, strong, and working cohesively

RESTRICTIONS
• Those with shoulder problems should avoid this exercise.

❷ Keeping your torso still, slide your hands together and then slide them back to the starting position. Complete two sets of 15 repetitions.

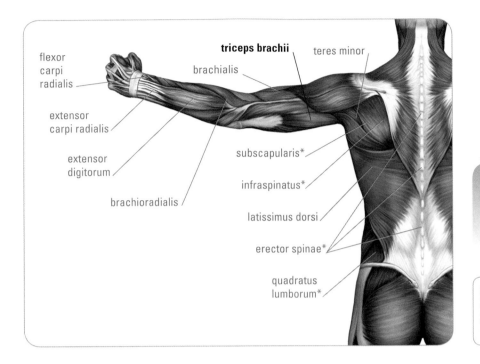

flexor
carpi
radialis

extensor
carpi radialis

extensor
digitorum

brachioradialis

triceps brachii

brachialis

teres minor

subscapularis*

infraspinatus*

latissimus dorsi

erector spinae*

quadratus
lumborum*

TARGET AREAS

Primary emphasis is on
the pectorals, deltoids,
and triceps.

ANNOTATION KEY

**Bold text indicates target
muscles**
Gray text indicates other
working muscles
* indicates deep muscles

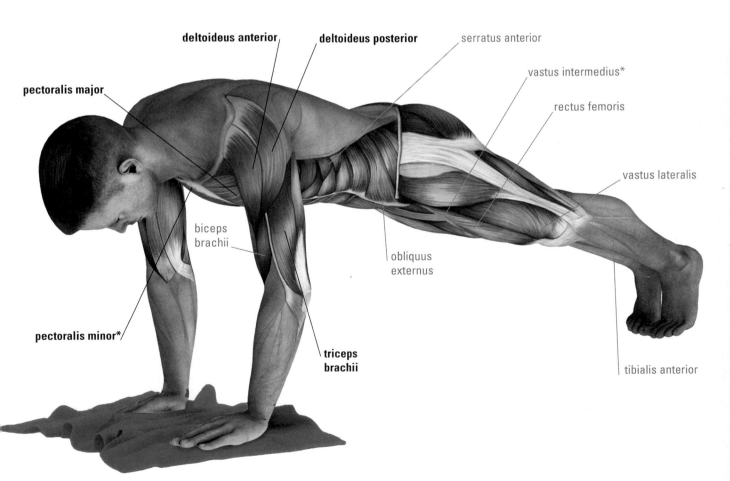

deltoideus anterior

deltoideus posterior

serratus anterior

vastus intermedius*

rectus femoris

vastus lateralis

pectoralis major

biceps
brachii

obliquus
externus

pectoralis minor*

**triceps
brachii**

tibialis anterior

SWISS BALL MEDICINE BALL PULLOVER

DO IT RIGHT

Be sure to ease into the stretch.

AVOID

Keeping your arms locked as you stretch behind the head.

1 Lie with your head and shoulders supported by a Swiss ball, with your feet on the floor, shoulder-width apart. Hold a medicine ball above your chest, with your arms fully extended.

LEVEL
• Intermediate

TIME
• 3-minute completion time

BENEFITS
• Helps keep the upper body stabilized and working cohesively

RESTRICTIONS
• Those with shoulder problems should avoid this exercise.

2 Bend your arms as necessary as you take the ball well behind your head, and then lengthen them as you raise them back into the starting position. Complete three sets of 15 repetitions.

pectoralis minor*

pectoralis major

serratus anterior

latissimus dorsi

triceps brachii

levator scapulae*

teres major

deltoideus posterior

TARGET AREAS

Primary emphasis is on the pectorals, deltoids, and triceps.

ANNOTATION KEY

Bold text indicates target muscles

Gray text indicates other working muscles

* indicates deep muscles

SWISS BALL PLANK WITH LEG LIFT

1 Position yourself on all fours, with a Swiss ball by your feet. Plant your hands on the ground with your arms fully extended, and place your left foot on top of the Swiss ball.

LEVEL
• Intermediate to advanced

TIME
• 2-minute completion time

BENEFITS
• Increases the ability to support your own body weight

RESTRICTIONS
• Pregnant women, though not restricted, should not undertake this exercise for longer than safely able.

MODIFICATION
• This exercise can be made easier by not raising your feet off the Swiss ball.

2 Place your right foot on top of the Swiss ball, too, and extend your legs fully.

❸ Raise your right foot off the ball and remain suspended in the plank position for 30 seconds (building up to 60 seconds). Replace your right foot and repeat with the left.

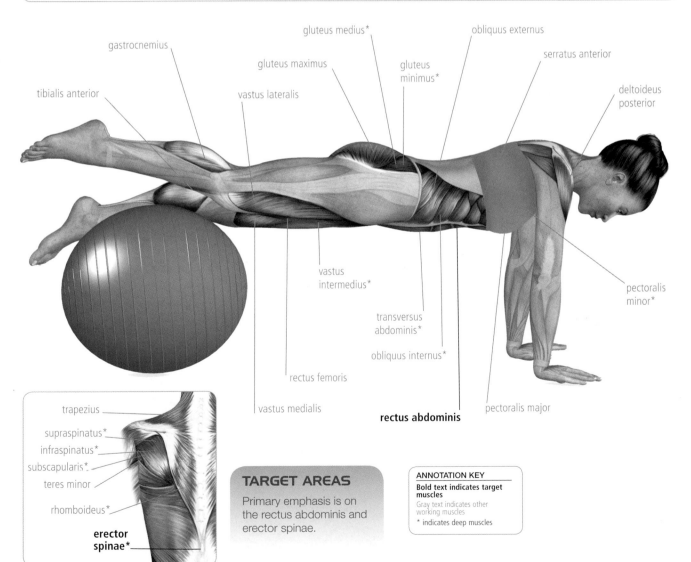

gluteus medius*

obliquus externus

gastrocnemius

gluteus maximus

gluteus minimus*

serratus anterior

tibialis anterior

vastus lateralis

deltoideus posterior

vastus intermedius*

pectoralis minor*

transversus abdominis*

obliquus internus*

rectus femoris

pectoralis major

vastus medialis

rectus abdominis

trapezius

supraspinatus*

infraspinatus*

subscapularis*

teres minor

rhomboideus*

erector spinae*

TARGET AREAS

Primary emphasis is on the rectus abdominis and erector spinae.

ANNOTATION KEY
Bold text indicates target muscles
Gray text indicates other working muscles
* indicates deep muscles

BODY SAW

1 Begin facedown on your forearms and toes.

2 Rock your body forward and then backward for three sets of 10 repetitions (working up to 20).

LEVEL
• Intermediate

TIME
• 3-minute completion time

BENEFITS
• Improves core strength and definition

RESTRICTIONS
• Those with lower-back problems should avoid this exercise.

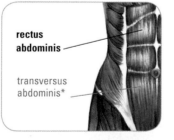

rectus
abdominis

transversus
abdominis*

TARGET AREAS

Primary emphasis is on the rectus abdominis and lower back.

ANNOTATION KEY
Bold text indicates target muscles
Gray text indicates other working muscles
* indicates deep muscles

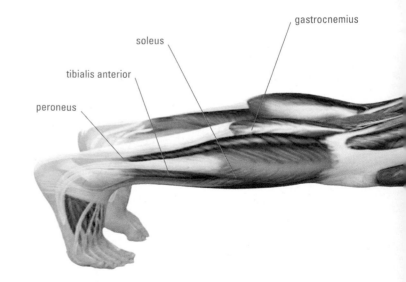

gastrocnemius

soleus

tibialis anterior

peroneus

DO IT RIGHT

Keep your body fully extended and in one straight line.

AVOID

Overusing your lower back by rising higher than parallel to the ground.

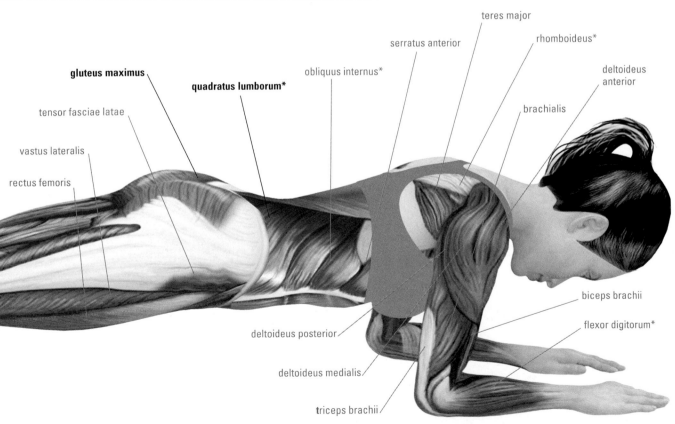

teres major

serratus anterior

rhomboideus*

obliquus internus*

deltoideus anterior

gluteus maximus

quadratus lumborum*

brachialis

tensor fasciae latae

vastus lateralis

rectus femoris

biceps brachii

flexor digitorum*

deltoideus posterior

deltoideus medialis

triceps brachii

BIG CIRCLES WITH MEDICINE BALL

❶ Stand holding a medicine ball out to your left side at arms' length, twisting your torso slightly toward the same direction.

Keep your torso straight throughout the exercise.

Performing the rotation too quickly.

LEVEL
• Intermediate

TIME
• 3-minute completion time

BENEFITS
• Improves abdominal range and definition

RESTRICTIONS
• Those with lower-back problems should avoid this exercise.

❷ In a controlled manner, swing your arms downward.

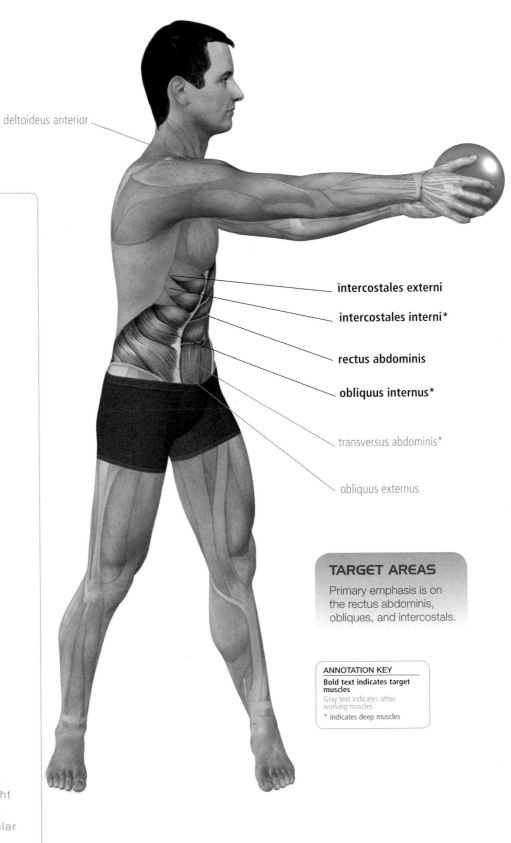

deltoideus anterior

intercostales externi

intercostales interni*

rectus abdominis

obliquus internus*

transversus abdominis*

obliquus externus

3 Continue carrying the ball smoothly through to your right side and over your head in a continuous 360-degree circular motion. Complete 30 circles, then repeat in the opposite direction for another 30.

TARGET AREAS

Primary emphasis is on the rectus abdominis, obliques, and intercostals.

ANNOTATION KEY
Bold text indicates target muscles
Gray text indicates other working muscles
* indicates deep muscles

McGILL CURL-UP

❶ Lie on your back with your right leg fully extended and your left leg bent. Place both hands palm-down underneath your lower back.

❷ Keeping your abdominal muscles braced, contract them slightly, bringing your head and shoulders off the floor. Hold for a 5-second count. Lower, and repeat for 10 repetitions, then switch legs.

LEVEL
• Intermediate

TIME
• 2-minute completion time

BENEFITS
• Helps strengthen the rectus abdominis and minimizes spinal compression

RESTRICTIONS
• Those with lower-back problems should avoid this exercise.

DO IT RIGHT	AVOID
Raise only your head and upper shoulders during this exercise.	Flattening your lower back to the ground.

108

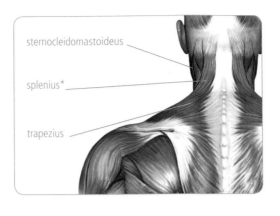

sternocleidomastoideus

splenius*

trapezius

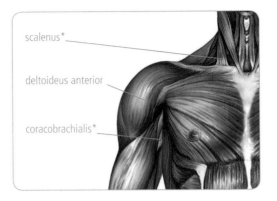

scalenus*

deltoideus anterior

coracobrachialis*

TARGET AREAS

Primary emphasis is on the rectus abdominis.

ANNOTATION KEY
Bold text indicates target muscles
Gray text indicates other working muscles
* indicates deep muscles

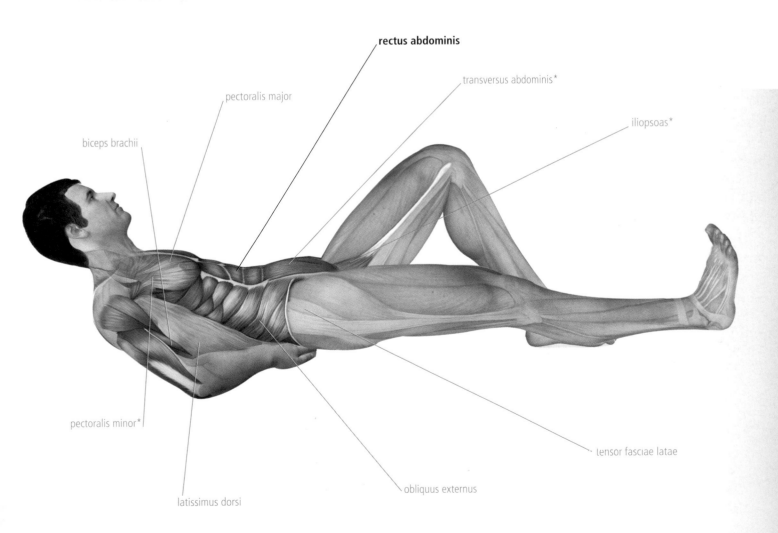

rectus abdominis

transversus abdominis*

pectoralis major

iliopsoas*

biceps brachii

pectoralis minor*

tensor fasciae latae

latissimus dorsi

obliquus externus

HIP CIRCLES

1 Sit on a Swiss ball with your hands on your hips and your feet close together.

3 Once you have performed 10 counterclockwise rotations, complete another 10 in a clockwise direction.

LEVEL
• Intermediate

TIME
• 3-minute completion time

BENEFITS
• Improves core stabilization and lower-back flexibility

RESTRICTIONS
• Those with back pain should avoid this exercise.

2 Keeping your core braced, use your pelvis to rotate the ball in small circular motions in a counterclockwise direction. Complete 10 circles.

TARGET AREAS

Primary emphasis is on the core and hips.

ANNOTATION KEY
Bold text indicates target muscles
Gray text indicates other working muscles
* indicates deep muscles

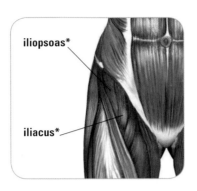

iliopsoas*

iliacus*

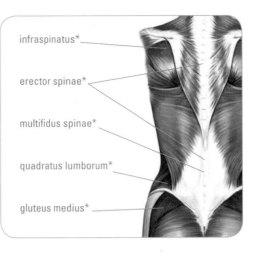

infraspinatus*

erector spinae*

multifidus spinae*

quadratus lumborum*

gluteus medius*

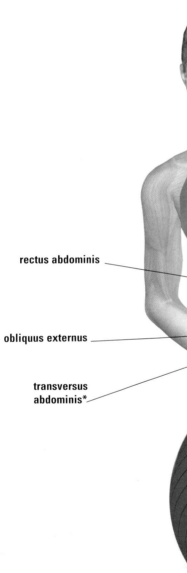

rectus abdominis

obliquus externus

transversus abdominis*

REVERSE BRIDGE ROTATION

1 Sit on a Swiss ball with your feet on the floor, shoulder-width apart, and a medicine ball in your hands.

2 Walk forward slowly while rolling your body along the Swiss ball until your upper back is supported by the ball. Extend your arms fully, holding the medicine ball directly above your chest.

LEVEL
• Intermediate

TIME
• 3-minute completion time

BENEFITS
• Will improve core strength

RESTRICTIONS
• Those with lower-back pain should avoid this exercise.

MODIFICATION
• This exercise can be modified by performing all the repetitions on one side before switching to the other.

3 Rotate your upper body toward the left and up onto your left shoulder.

4 Slowly return to the center position and continue the movement over toward the right. Complete three sets of 15 repetitions per side.

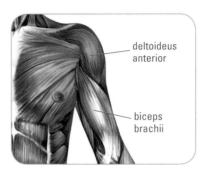

deltoideus anterior

biceps brachii

TARGET AREAS

Primary emphasis is on the rectus abdominis and oblique muscles.

ANNOTATION KEY
Bold text indicates target muscles
Gray text indicates other working muscles
* indicates deep muscles

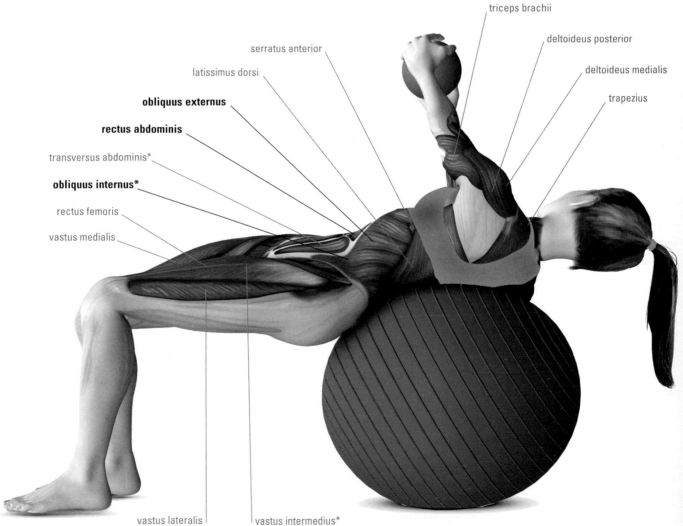

triceps brachii

deltoideus posterior

deltoideus medialis

trapezius

serratus anterior

latissimus dorsi

obliquus externus

rectus abdominis

transversus abdominis*

obliquus internus*

rectus femoris

vastus medialis

vastus lateralis

vastus intermedius*

PLANK KNEE PULL-IN

1 Begin by assuming a standard plank position.

DO IT RIGHT	AVOID
Keep your body in a straight line throughout the exercise.	Bending the knee of the supporting leg.

2 Draw your left knee into your chest while leaning forward and flexing your foot. Your right leg should be up on its toes.

LEVEL
• Advanced

TIME
• 2-minute completion time

BENEFITS
• Improves core stabilization and flexibility

RESTRICTIONS
• Those with lower-back or wrist pain should avoid this exercise.

MODIFICATION
• This exercise can be made easier by using a wall to help support the raised leg.

3 Extend your right leg through the heel, and rock your body back, shifting your weight into your left foot.

4 Drop your head between your arms, and straighten and raise your left leg toward the ceiling. Repeat the entire exercise 10 times per leg.

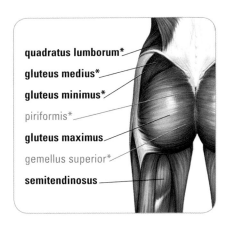

quadratus lumborum*
gluteus medius*
gluteus minimus*
piriformis*
gluteus maximus
gemellus superior*
semitendinosus

TARGET AREAS

Primary emphasis is on the core, hamstrings, glutes, and scapular area.

ANNOTATION KEY
Bold text indicates target muscles
Gray text indicates other working muscles
* indicates deep muscles

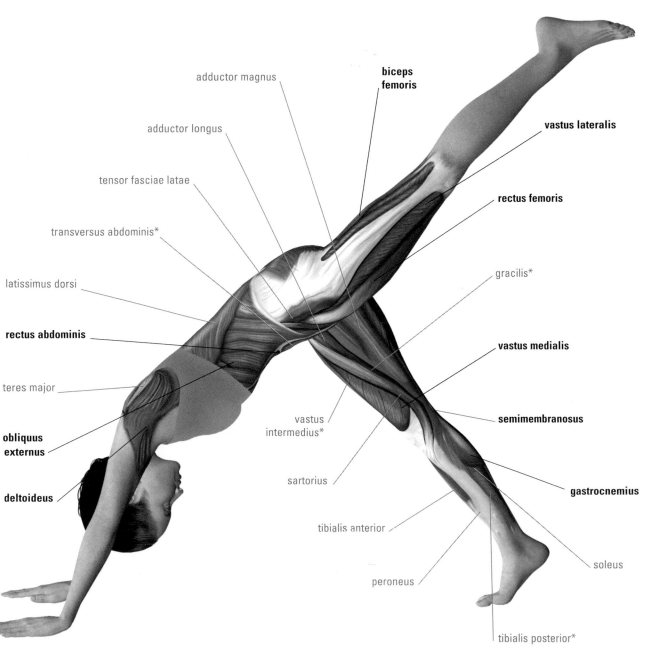

adductor magnus

biceps femoris

adductor longus

vastus lateralis

tensor fasciae latae

rectus femoris

transversus abdominis*

gracilis*

latissimus dorsi

rectus abdominis

vastus medialis

teres major

semimembranosus

obliquus externus

vastus intermedius*

deltoideus

sartorius

gastrocnemius

tibialis anterior

soleus

peroneus

tibialis posterior*

ABDOMINAL HIP LIFT

DYNAMIC

① Lie on your back with your arms at your sides and your legs as high in the air as possible, crossed at the ankles.

DO IT RIGHT

Keep your legs as straight as possible throughout the exercise.

AVOID

Using momentum, and bringing the lower back too much into play.

LEVEL
• Intermediate

TIME
• 2-minute completion time

BENEFITS
• Improves core strength

RESTRICTIONS
• Those with lower-back problems should avoid this exercise.

MODIFICATION
• This exercise can be modified by bending the legs to reduce the stress on the abdominals.

② Keep your legs tightly pressed together and your glutes activated. Push through your triceps to lift your hips upward. Lower slowly. Perform 10 repetitions, then cross your legs the other way and perform another 10.

116

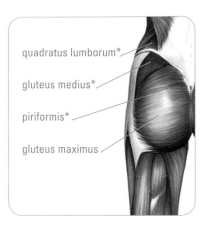

quadratus lumborum*

gluteus medius*

piriformis*

gluteus maximus

ANNOTATION KEY
Bold text indicates target muscles
Gray text indicates other working muscles
* indicates deep muscles

TARGET AREAS

Primary emphasis is the rectus abdominis, with secondary assistance from the triceps.

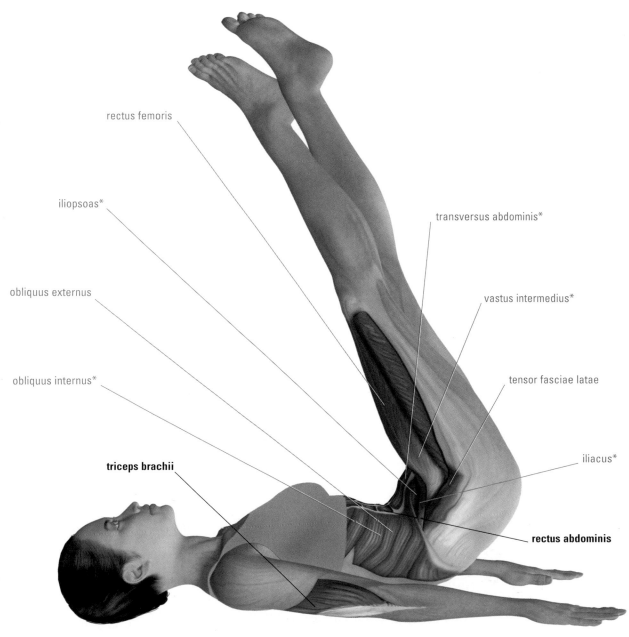

rectus femoris

iliopsoas*

obliquus externus

obliquus internus*

triceps brachii

transversus abdominis*

vastus intermedius*

tensor fasciae latae

iliacus*

rectus abdominis

STRAIGHT LEG RAISE

DYNAMIC

1 Lie on your back with your arms at your sides, parallel to your body. Lift your legs slightly off the floor. (You can start higher, but no more than about 45 degrees.)

LEVEL
• Intermediate

TIME
• 2-minute completion time

BENEFITS
• Improves core strength and support

RESTRICTIONS
• Those with lower-back problems should avoid this exercise.

MODIFICATION
• This exercise can be modified by bending the legs to reduce the stress on the abdominals.

DO IT RIGHT

Keep your upper body braced.

AVOID

Using momentum, and bringing the lower back too much into play.

2 Raise both legs until they are nearly vertical to the floor, then lower them just short of the floor. Complete two sets of 20 repetitions.

ANNOTATION KEY
Bold text indicates target muscles
Gray text indicates other working muscles
* indicates deep muscles

TARGET AREAS

Primary emphasis is on the rectus abdominis.

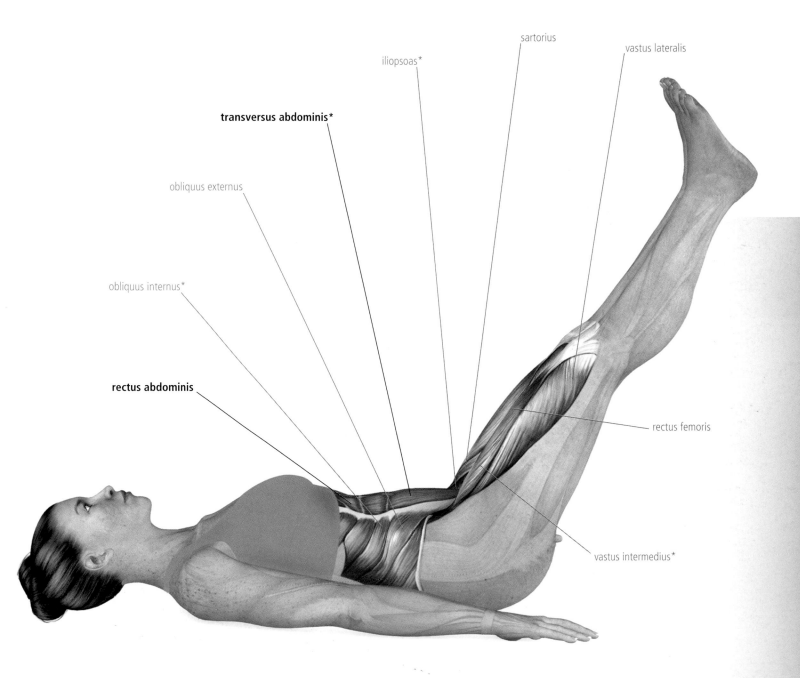

iliopsoas*

sartorius

vastus lateralis

transversus abdominis*

obliquus externus

obliquus internus*

rectus abdominis

rectus femoris

vastus intermedius*

SEATED RUSSIAN TWIST

1 Sit upright with your legs bent and your feet flat on the floor. Extend your arms straight ahead, and lean back slightly to activate your core.

DO IT RIGHT
Twist smoothly and with control, keeping your back flat.

AVOID
Shifting your feet or knees as you twist.

LEVEL
• Intermediate

TIME
• 2-minute completion time

BENEFITS
• Stabilizes and strengthens the core

RESTRICTIONS
• Those with lower-back problems should avoid this exercise.

2 In a smooth motion, rotate your upper body to the side, and then return to the center. Repeat the rotation on the other side.

3 Return to the center, and repeat the full twist, performing three sets of 20 repetitions.

120

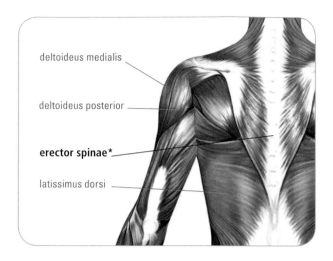

deltoideus medialis

deltoideus posterior

erector spinae*

latissimus dorsi

TARGET AREAS

Primary emphasis is on the erector spinae, obliques, rectus abdominis, and transversus abdominis.

ANNOTATION KEY

Bold text indicates target muscles

Gray text indicates other working muscles

* indicates deep muscles

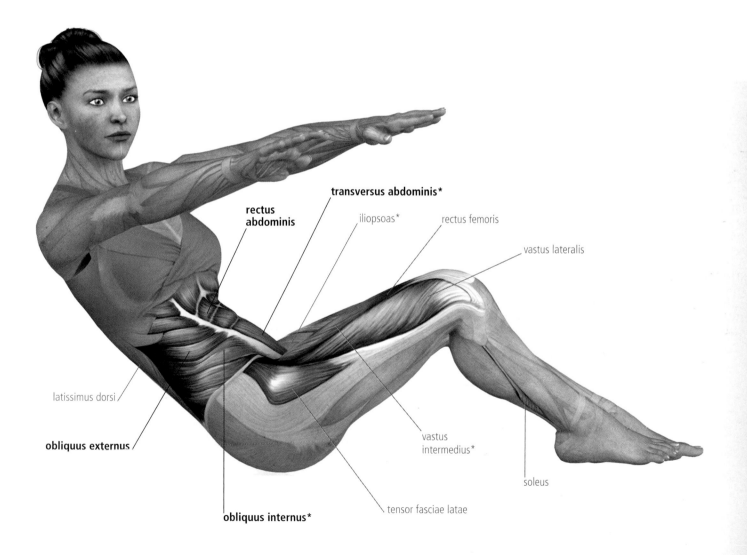

transversus abdominis*

rectus abdominis

iliopsoas*

rectus femoris

vastus lateralis

latissimus dorsi

obliquus externus

vastus intermedius*

soleus

obliquus internus*

tensor fasciae latae

SINGLE-LEG CIRCLES

DYNAMIC

❶ Lie on your back with your arms at your sides, parallel to your body. Raise your right leg as high as you are able, with your foot pointing slightly outward.

❶ Lie on your back with your arms at your sides, parallel to your body. Raise your right leg as high as you are able, with your foot pointing slightly outward.

DO IT RIGHT
Keep your hips still throughout the exercise.

AVOID
Performing the circles at excessive speed.

LEVEL
• Beginner

TIME
• 2-minute completion time

BENEFITS
• Improves abdominal and pelvic strength

RESTRICTIONS
• Those with lower-back problems should avoid this exercise.

❷ Cross your right leg up and over your body as if tracing a circle. Do this five times in a clockwise direction, then another five times counterclockwise. Switch legs, and repeat the exercise.

122

TARGET AREAS

Primary emphasis is on the abdominals, hips, hamstrings, and your inner and outer thighs.

ANNOTATION KEY
Bold text indicates target muscles
Gray text indicates other working muscles
* indicates deep muscles

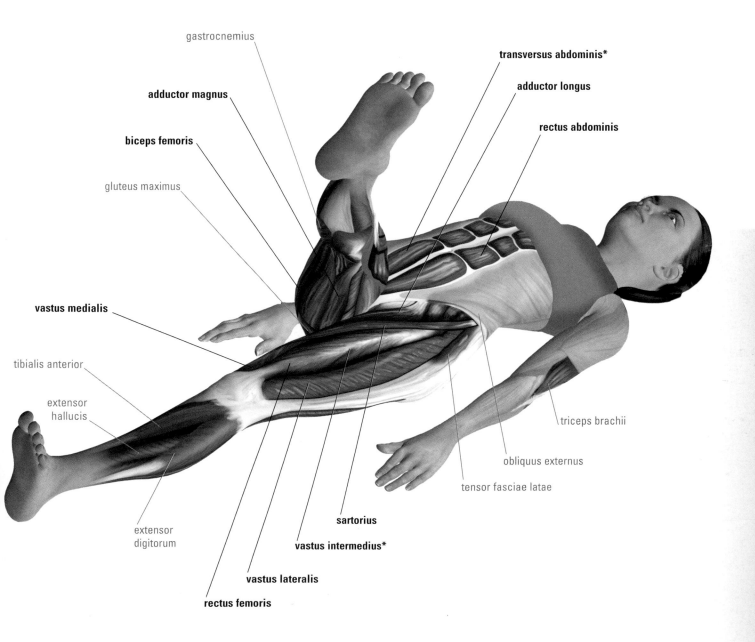

gastrocnemius

transversus abdominis*

adductor magnus

adductor longus

biceps femoris

rectus abdominis

gluteus maximus

vastus medialis

tibialis anterior

extensor hallucis

triceps brachii

obliquus externus

tensor fasciae latae

extensor digitorum

sartorius

vastus intermedius*

vastus lateralis

rectus femoris

PRONE HEEL-BEATS

DYNAMIC

❶ Lie facedown with your arms slightly elevated at your sides. Lift your legs a little way off the floor.

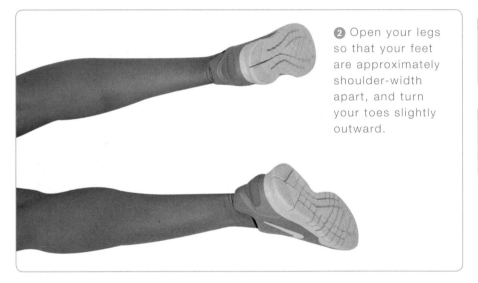

❷ Open your legs so that your feet are approximately shoulder-width apart, and turn your toes slightly outward.

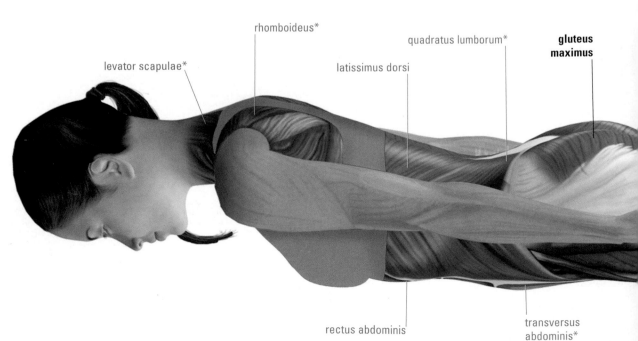

LEVEL
• Intermediate

TIME
• 2-minute completion time

BENEFITS
• Targets the core stabilizers

RESTRICTIONS
• Those with lower-back problems should avoid this exercise.

DO IT RIGHT
Keep your legs elevated throughout the exercise.

AVOID
Overengaging your shoulders.

levator scapulae*

rhomboideus*

quadratus lumborum*

latissimus dorsi

gluteus maximus

rectus abdominis

transversus abdominis*

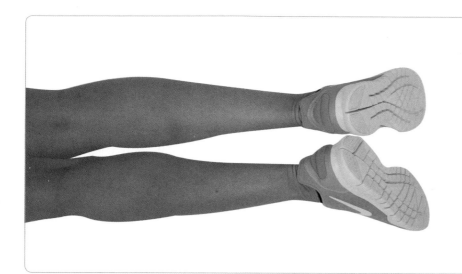

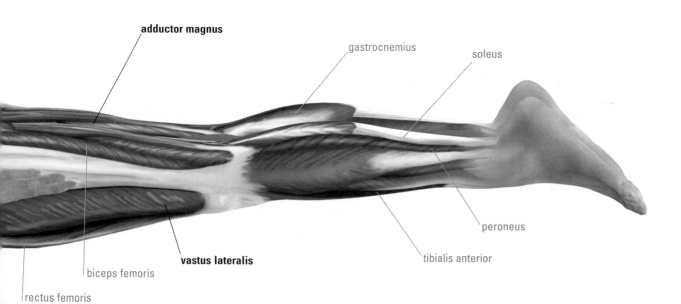

3 Beat your heels together 10 times, then rest. Complete three sets of 10 repetitions.

splenius*

trapezius

deltoideus posterior

teres minor

teres major

triceps brachii

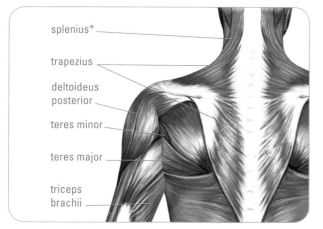

TARGET AREAS

Primary emphasis is on the glutes, back, and inner thighs.

ANNOTATION KEY

Bold text indicates target muscles

Gray text indicates other working muscles

* indicates deep muscles

---

adductor magnus

gastrocnemius

soleus

peroneus

vastus lateralis

tibialis anterior

biceps femoris

rectus femoris

CLAMSHELL SERIES

1 Position yourself on your right hip, placing your right forearm on the floor to support yourself. Put your left hand on your left hip. Keep your legs slightly bent, lying one on top of the other.

DO IT RIGHT
Keep your spine straight throughout the exercise.

AVOID
Letting your hips rise while lifting your knees.

2 Keeping a straight spine, your right leg on the floor, and your feet together, lift your left knee 10 times.

LEVEL
• Advanced

TIME
• 6-minute completion time

BENEFITS
• Increases pelvic stability and strengthens the abductor muscles

RESTRICTIONS
• Those with lower-back or shoulder problems should avoid this exercise.

3 Next, holding your knees and feet together, lift your feet off the floor.

4 While your feet are raised, open and close your knees 10 times, again moving only your left leg.

5 Finish step 4 with your knees open, then lift your left leg and straighten it, without moving the thigh. Bend the leg again. Do this 10 times, then switch sides to repeat the entire series.

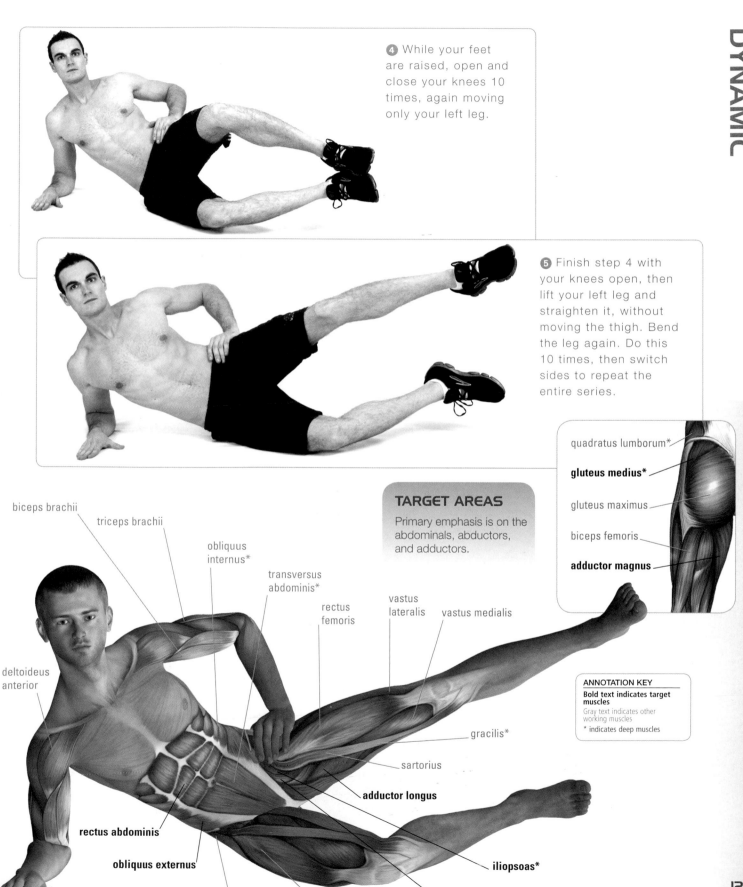

TARGET AREAS

Primary emphasis is on the abdominals, abductors, and adductors.

quadratus lumborum*

gluteus medius*

gluteus maximus

biceps femoris

adductor magnus

biceps brachii

triceps brachii

obliquus internus*

transversus abdominis*

rectus femoris

vastus lateralis

vastus medialis

deltoideus anterior

gracilis*

sartorius

adductor longus

rectus abdominis

obliquus externus

iliopsoas*

tensor fasciae latae

vastus intermedius*

iliacus*

ANNOTATION KEY
Bold text indicates target muscles
Gray text indicates other working muscles
* indicates deep muscles

127

SCISSORS

❶ Lie on your back with your arms at your sides, parallel to your body, and your legs bent at a 90-degree angle and off the ground.

DO IT RIGHT

Keep your pelvis stabilized and your spine straight.

AVOID

Overextending your raised leg.

LEVEL
• Intermediate

TIME
• 2-minute completion time

BENEFITS
• Improves core stability, with increased abdominal strength and endurance

RESTRICTIONS
• Those with tight hamstrings should avoid this exercise.

MODIFICATION
• This exercise can be modified by performing all the repetitions on one leg before switching to the other.

❷ Lower your right leg and straighten your left leg up toward your trunk, bracing your calf with your hands and contracting your abdominal muscles at the same time. Hold for 5 seconds.

3 Bring both legs back to the starting position, and repeat step 2, this time with your left leg down and your right leg up. Complete 10 repetitions per leg.

ANNOTATION KEY
Bold text indicates target muscles
Gray text indicates other working muscles
* indicates deep muscles

TARGET AREAS

Primary emphasis is on the rectus femoris, biceps femoris, and rectus abdominis.

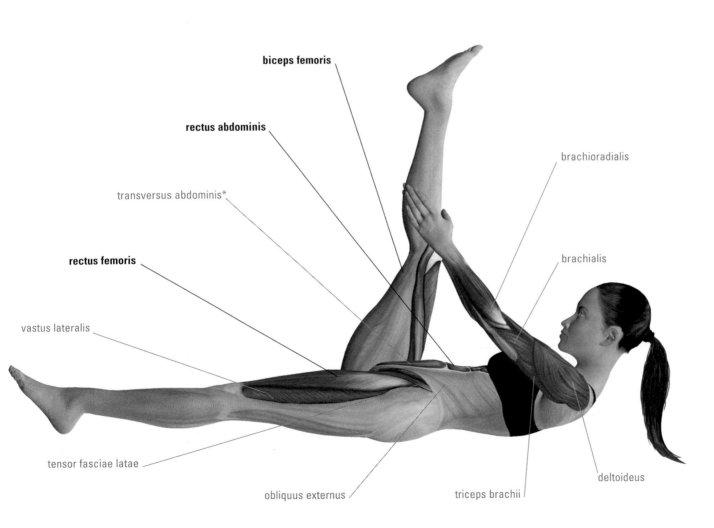

biceps femoris

rectus abdominis

transversus abdominis*

rectus femoris

vastus lateralis

tensor fasciae latae

obliquus externus

brachioradialis

brachialis

triceps brachii

deltoideus

MEDICINE BALL AB CURL

1 Lie faceup on a Swiss ball with your back supported by the ball. Hold a medicine ball in both hands at arms' length above your chest.

2 Raise your neck and shoulders off the ball while contracting your abdominals, then lower yourself back down. Complete three sets of 15 repetitions.

LEVEL
- Intermediate

TIME
- 3-minute completion time

BENEFITS
- Improves core strength

RESTRICTIONS
- Those with lower-back problems should avoid this exercise.

MODIFICATION
- This exercise can be made easier by performing it without a medicine ball.

TARGET AREAS

Primary emphasis is on the rectus abdominis.

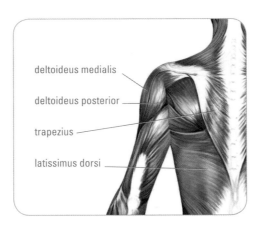

deltoideus medialis

deltoideus posterior

trapezius

latissimus dorsi

ANNOTATION KEY

Bold text indicates target muscles

Gray text indicates other working muscles

* indicates deep muscles

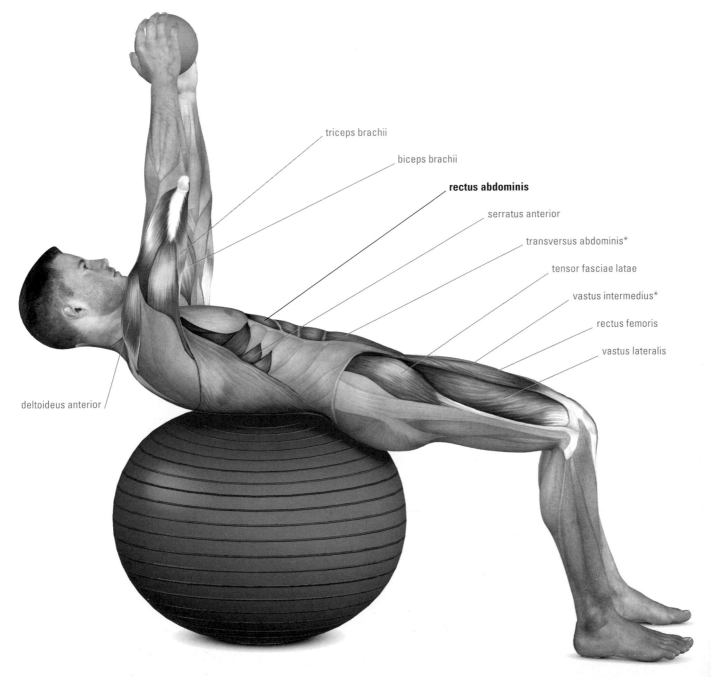

triceps brachii

biceps brachii

rectus abdominis

serratus anterior

transversus abdominis*

tensor fasciae latae

vastus intermedius*

rectus femoris

vastus lateralis

deltoideus anterior

SIT-UP AND THROW

DYNAMIC

① Lie on your back with your legs slightly bent and your feet flat on the floor. Holding a medicine ball in your hands, stretch your arms out behind your head.

DO IT RIGHT

Lead with your abdominals as if a string was hoisting you up from your belly button.

AVOID

Overusing your neck.

② Push through your heels for support, and bring your arms forward while lifting your torso off the ground and contracting your abdominals. Throw the ball to your partner, receive it back, and lower yourself back down. Perform two sets of 15 repetitions.

LEVEL
• Beginner

TIME
• 2-minute completion time

BENEFITS
• Improves strength and definition of the abdominals

RESTRICTIONS
• Those with lower-back problems should avoid this exercise.

NOTE
• This exercise requires the assistance of a partner.

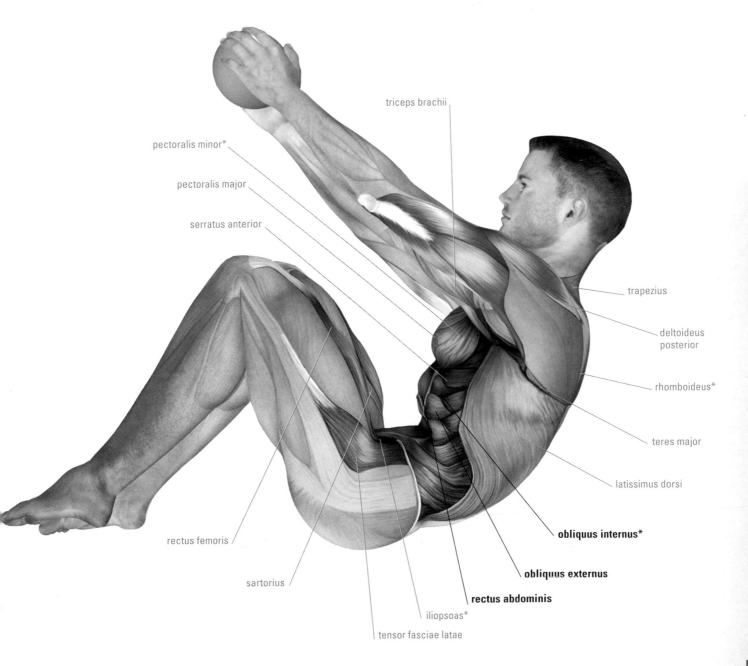

TARGET AREAS

Primary emphasis is on the
rectus abdominis muscles.

ANNOTATION KEY
**Bold text indicates target
muscles**
Gray text indicates other
working muscles
* indicates deep muscles

triceps brachii

pectoralis minor*

pectoralis major

serratus anterior

trapezius

deltoideus
posterior

rhomboideus*

teres major

latissimus dorsi

rectus femoris

sartorius

obliquus internus*

obliquus externus

rectus abdominis

iliopsoas*

tensor fasciae latae

BICYCLE CRUNCH

DO IT RIGHT	AVOID
Keep your chin off your chest, and keep both hips on the floor.	Pulling with your hands or arching your back.

❶ Lie supine on the floor with your knees bent. Bring your hands behind your head, lifting your legs off the floor.

❷ Roll up with your torso, reaching your left elbow to your right knee while extending the left leg in front of you. Imagine pulling your shoulder blades off the floor and twisting from your ribs and oblique muscles.

LEVEL
• Advanced

TIME
• 2-minute completion time

BENEFITS
• Stabilizes the core and strengthens the abdominals

RESTRICTIONS
• Those with lower-back or neck problems should avoid this exercise.

❸ Switch sides. Complete the movement six times on each side.

MODIFICATION

For this easier variant, begin with both feet on the floor. Place your left ankle on your right thigh, near the knee. Reach your right elbow toward your left knee. Complete six repetitions on each side.

TARGET AREAS

Primary emphasis is on the abdominal muscles and obliques.

ANNOTATION KEY
Bold text indicates target muscles
Gray text indicates other working muscles
* indicates deep muscles

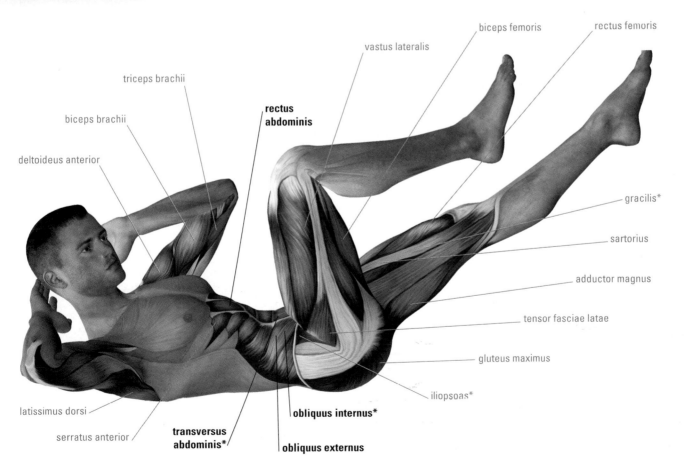

- vastus lateralis
- biceps femoris
- rectus femoris
- triceps brachii
- **rectus abdominis**
- biceps brachii
- deltoideus anterior
- gracilis*
- sartorius
- adductor magnus
- tensor fasciae latae
- gluteus maximus
- iliopsoas*
- latissimus dorsi
- serratus anterior
- **transversus abdominis***
- **obliquus internus***
- **obliquus externus**

STEP-DOWN

① Stand on a block with left foot very close to the left edge and the right one dangling off the block. Hold your arms out straight in front of you.

DO IT RIGHT
Push through your planted heel.

AVOID
Letting your knee extend past your toes.

② Bend your left leg at the knee, lowering yourself toward the ground as your right leg dangles below the top of the block.

③ Push through your left heel to return to the starting position. Complete two sets of 15 repetitions per leg.

LEVEL
• Advanced

TIME
• 2-minute completion time

BENEFITS
• Strengthens the pelvic and knee stabilizers

RESTRICTIONS
• Those with knee problems should avoid this exercise.

MODIFICATION
• This exercise can be made easier by pressing your palms against a wall for support.

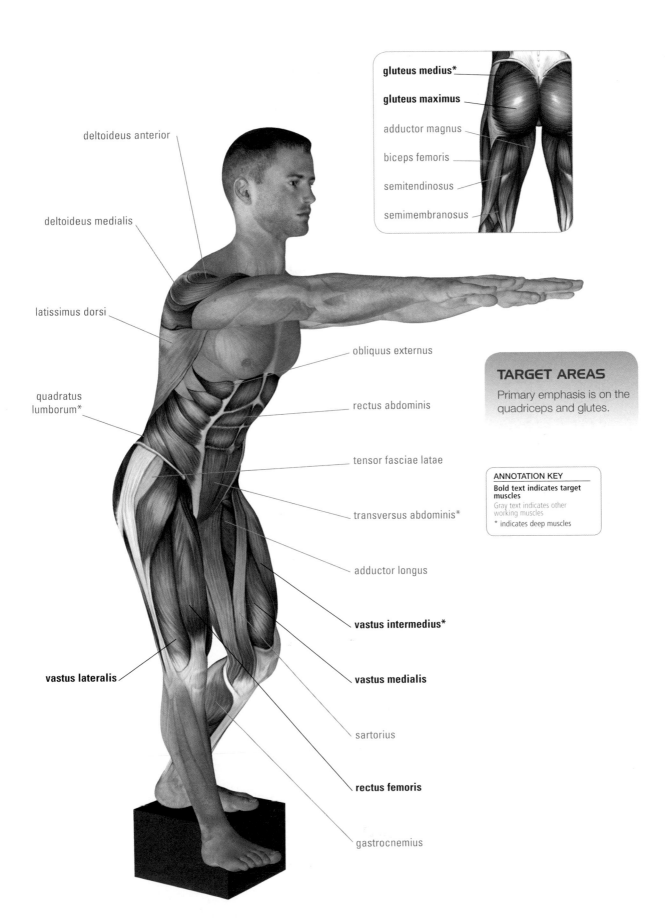

gluteus medius*

gluteus maximus

adductor magnus

biceps femoris

semitendinosus

semimembranosus

deltoideus anterior

deltoideus medialis

latissimus dorsi

quadratus lumborum*

vastus lateralis

obliquus externus

rectus abdominis

tensor fasciae latae

transversus abdominis*

adductor longus

vastus intermedius*

vastus medialis

sartorius

rectus femoris

gastrocnemius

TARGET AREAS
Primary emphasis is on the quadriceps and glutes.

ANNOTATION KEY

Bold text indicates target muscles

Gray text indicates other working muscles

* indicates deep muscles

SPINE TWIST

① Sit on the ground with your legs extended and your feet slightly wider apart than your hips. Hold your back straight, and raise your arms out to the sides, fully extended, at 90 degrees to your torso.

LEVEL
• Intermediate

TIME
• 1-minute completion time

BENEFITS
• Improves back flexibility, and strengthens and lengthens the torso

RESTRICTIONS
• Those with lower-back problems should avoid this exercise.

② Keeping your abdominals pulled in, twist your waist to the left, taking your entire upper body with it, then return to the central position.

3 Repeat the movement, this time turning to the right. Complete three twists in each direction.

TARGET AREAS

Primary emphasis is on the entire core, latissimus dorsi, deltoids, glutes, and hamstrings.

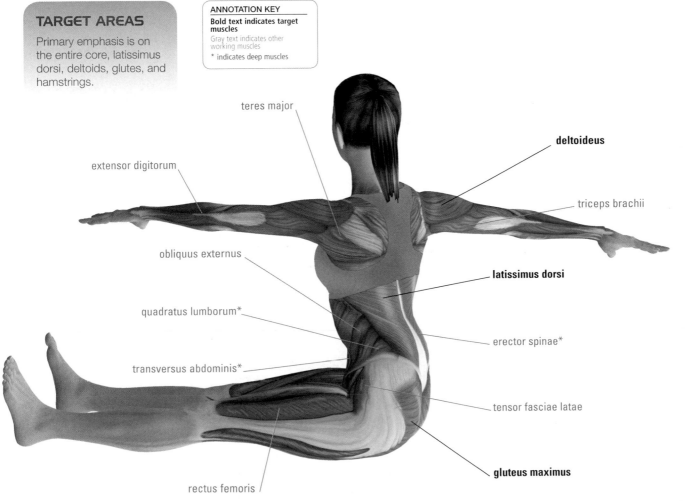

teres major

extensor digitorum

obliquus externus

quadratus lumborum*

transversus abdominis*

rectus femoris

deltoideus

triceps brachii

latissimus dorsi

erector spinae*

tensor fasciae latae

gluteus maximus

STIFF-LEGGED DEADLIFT

1 Holding a pair of dumbbells loosely against your outer thighs, stand with your feet shoulder-width apart, your knees slightly bent, and your rear pushed out slightly.

DO IT RIGHT
Keep your back flat at all times.

AVOID
Straining your lower back excessively.

LEVEL
• Intermediate

TIME
• 3-minute completion time

BENEFITS
• Improves lower-body flexibility and stabilization

RESTRICTIONS
• Those with lower-back problems should avoid this exercise.

2 Maintaining a flat back at all times, bring the dumbbells around in front of you and lower them toward the floor, feeling the main stretch in the back of the legs.

3 Return to the starting position, and complete three sets of 15 repetitions.

140

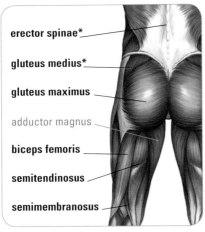

erector spinae*
gluteus medius*
gluteus maximus
adductor magnus
biceps femoris
semitendinosus
semimembranosus

TARGET AREAS

Primary emphasis is on the glutes, hamstrings, and erector spinae.

ANNOTATION KEY
Bold text indicates target muscles
Gray text indicates other working muscles
* indicates deep muscles

levator scapulae*

trapezius

rhomboideus*

gluteus maximus

latissimus dorsi

rectus abdominis

SWIMMING

❶ Lie on your stomach with your arms stretched out in front of you and your legs stretched out behind. Raise your left arm and right leg off the floor at the same time, along with your head and shoulders, then lower them all back down.

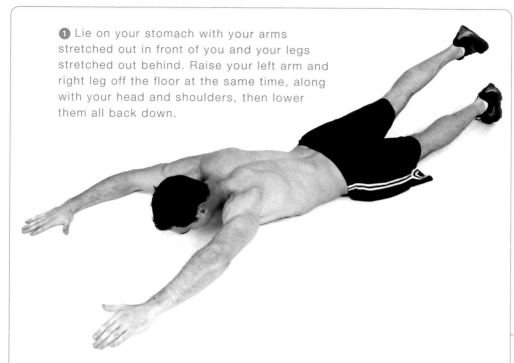

DO IT RIGHT

Raise your arms and legs as high as possible.

AVOID

Overstressing the neck.

LEVEL
• Intermediate

TIME
• 1-minute completion time

BENEFITS
• Improves lower-back strength and support

RESTRICTIONS
• Those with lower-back problems should avoid this exercise.

❷ Repeat the exercise with your opposite limbs. Complete 10 repetitions per side.

MODIFICATION

This exercise can be made more difficult by raising both arms and legs at the same time.

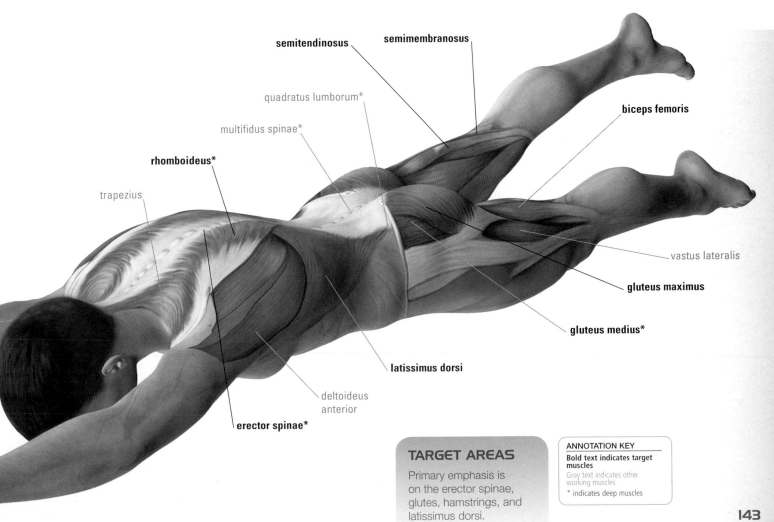

semitendinosus

semimembranosus

quadratus lumborum*

biceps femoris

multifidus spinae*

rhomboideus*

trapezius

vastus lateralis

gluteus maximus

gluteus medius*

latissimus dorsi

deltoideus anterior

erector spinae*

TARGET AREAS

Primary emphasis is on the erector spinae, glutes, hamstrings, and latissimus dorsi.

ANNOTATION KEY
Bold text indicates target muscles
Gray text indicates other working muscles
* indicates deep muscles

WORKOUTS

The following theme-based workouts will help you make the most of your core stability. Remember always to commit yourself fully to a given exercise rather than rushing through it. Allow yourself time to make each movement more effective, as well as to require less assistance from local ancillary muscles. Challenge yourself further by executing each movement to the best of your ability rather than attempting to complete a sequence more quickly. Above all, have fun.

BEGINNER WORKOUT

Although this sequence is suitable for all levels, it is especially aimed at those who are new to core training.

1 Plank, page 40

2 Side Plank, page 42

3 Quadruped, page 50

4 Bridge, page 56

5 Scissors, page 128

6 Body Saw, page 104

7 Hip Circles, page 110

8 Straight Leg Raise, page 118

9 Big Circles with Medicine Ball, page 106

10 Side-Lying Hip Abduction, page 76

LATERAL-CORE WORKOUT

WORKOUTS

The emphasis in this routine is on stabilizing, strengthening, and increasing the definition of the abdominal region.

❶ Side Plank, page 42

❷ Transverse Abs, page 52

❸ Rotated Back Extension, page 74

❹ Lateral Roll, page 70

❺ Bicycle Crunch, page 134

6 Spine Twist, page 138

7 Push-Up Hand Walkover, page 94

8 Swiss Ball Hip Crossover, page 88

9 Swiss Ball Walk-Around, page 90

10 Big Circles with Medicine Ball, page 106

ANTERIOR WORKOUT

In this routine, the focus is on stabilizing, strengthening, and showcasing the rectus abdominis.

1 Body Saw, page 104

2 Bottom Push-Up Hold, page 60

3 Front Plank, page 44

4 Swiss Ball Roll-Out, page 84

5 Swiss Ball Jackknife, page 86

6 Seated Pelvic Tilt, page 31

❼ Sit-Up and Throw, page 132

❽ Transverse Abs, page 52

❾ Swiss Ball Medicine Ball Pullover, page 100

❿ Plank Knee Pull-In, page 114

SPORTS WORKOUT

This sequence is all about readying the core for functional athletic performance.

1 Seated Russian Twist, page 120

2 Swiss Ball Jackknife, page 86

3 Sit-Up and Throw, page 132

4 Swiss Ball Plank with Leg Lift, page 102

5 Step-Down, page 136

6 Plank Roll-Down, page 46

7 Lateral Roll, page 70

8 Inverted Hamstring, page 66

9 Swiss Ball Hyperextension, page 72

10 Towel Fly, page 98

DOWN-UNDER WORKOUT

This regimen works the lower part of the body, placing the emphasis on the legs and their role in core development and performance.

❶ Scissors, page 128

❷ Single-Leg Balance, page 62

❸ High Lunge, page 64

❹ Quadruped, page 50

❺ Static Sumo Squat, page 68

6 Inverted Hamstring, page 66

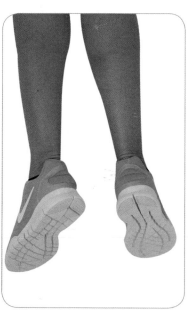

7 Stiff-Legged Deadlift, page 140

8 Single-Leg Circles, page 122

9 Clamshell Series, page 126

10 Prone Heel-Beats, page 124

155

SUICIDE STABILITY WORKOUT

This sequence is for the insatiable diehard who wishes to maximize core stability and strength and also relishes a challenge.

❶ Swiss Ball Plank with Leg Lift, page 102

❷ Step-Down, page 136

❸ High Lunge, page 64

❹ Swiss Ball Jackknife, page 86

❺ Push-Up Hand Walkover, page 94

❻ Chair Dip, page 96

7 Swiss Ball Walk-Around, page 90

8 Towel Fly, page 98

9 Thigh Rock-Back, page 38

10 Transverse Abs, page 52

GLOSSARY

GENERAL TERMS

abduction: Movement away from the body.

adduction: Movement toward the body.

anterior: Located in the front.

cardiovascular exercise: Any exercise that increases the heart rate, making oxygen and nutrient-rich blood available to working muscles.

cardiovascular system: The circulatory system that distributes blood throughout the body; it includes the heart, lungs, arteries, veins, and capillaries.

core: Refers to the deep muscle layers that lie close to the spine and provide structural support for the entire body. The core is divisible into major core and minor core. The major core muscles are on the trunk and include the belly area and the mid- and lower back. This area encompasses the pelvic-floor muscles (levator ani, pubococcygeus, iliococcygeus, pubo-rectalis, and coccygeus), the abdominals (rectus abdominis, transversus abdominis, obliquus externus, and obliquus internus), the spinal extensors (multifidus spinae, erector spinae, splenius, longissimus thoracis, and semispinalis), and the diaphragm. The minor core muscles include the latissimus dorsi, gluteus maximus, and trapezius (upper, middle, and lower). These minor core muscles assist the major muscles when the body engages in activities or movements that require added stability.

crunch: A common abdominal exercise that calls for curling the shoulders toward the pelvis while lying supine with hands behind the head and knees bent.

curl: An exercise movement, usually targeting the biceps brachii, that calls for a weight to be moved through an arc, in a "curling" motion.

deadlift: An exercise movement that calls for lifting a weight, such as a barbell, off the ground from a stabilized bent-over position.

dumbbell: A basic piece of equipment that consists of a short bar on which plates are secured. A person can use a dumbbell in one or both hands during an exercise. Most gyms offer dumbbells with the weight plates welded on and poundage indicated on the plates, but many dumbbells intended for home use come with removable plates that allow you to adjust the weight.

dynamic exercise: An exercise that includes movement through the joints and muscles.

extension: The act of straightening.

extensor muscle: A muscle serving to extend a body part away from the body.

flexion: The bending of a joint.

flexor muscle: A muscle that decreases the angle between two bones, such as bending the arm at the elbow or raising the thigh toward the stomach.

fly: An exercise movement in which the hand and arm move through an arc while the elbow is kept at a constant angle. A fly works the muscles of the upper body.

iliotibial band (ITB): A thick band of fibrous tissue that runs down the outside of the leg, beginning at the hip and extending to the outer side of the tibia, just below the knee joint. The ITB works in conjunction with several of the thigh muscles to provide stability to the outside of the knee joint.

lateral: Located on, or extending toward, the outside.

medial: Located on, or extending toward, the middle.

medicine ball: A small weighted ball used in weight training and toning.

neutral position (spine): A spinal position resembling an S shape, consisting of a lordosis (backward curvature) in the lower back, when viewed in profile.

posterior: Located behind.

press: An exercise movement that calls for moving a weight, or other resistance, away from the body.

range of motion: The distance and direction a joint can move between the flexed position and the extended position.

resistance band: Any rubber tubing or flat band device that provides a resistive force used for strength training. Also called a "fitness band," "stretching band," and "stretch tube."

rotator muscle: One of a group of muscles that assist the rotation of a joint, such as the hip or the shoulder.

scapula: The protrusion of bone on the mid- to upper back, also known as the "shoulder blade."

squat: An exercise that calls for moving the hips back and bending the knees and hips to lower the torso (and an accompanying weight, if desired) and then returning to the upright position. A squat primarily targets the muscles of the thighs, hips and buttocks, and hamstrings.

static exercise: An isometric form of exercise, without movement of the joints, that is held for a specific period of time.

Swiss ball: A flexible, inflatable PVC ball, measuring approximately 14 to 34 inches in circumference, used for weight training, physical therapy, balance training, and many other exercise regimens. It is also called a "balance ball," "fitness ball," "stability ball," "exercise ball," "gym ball," "physioball," "body ball," and many other names.

warm-up: Any form of light exercise of short duration that prepares the body for more intense activity.

weight: Refers to the plates or weight stacks, or the actual poundage listed on the bar or dumbbell.

LATIN TERMS

The following glossary explains the Latin terminology used to describe the body's musculature. Where words are derived from Greek, this is indicated.

CHEST

coracobrachialis: Greek *korakoeidés*, "ravenlike," and *brachium*, "arm"

pectoralis (major and minor): *pectus*, "breast"

ABDOMEN

obliquus externus: *obliquus*, "slanting," and *externus*, "outward"

obliquus internus: *obliquus*, "slanting," and *internus*, "within"

rectus abdominis: *rego*, "straight, upright," and *abdomen*, "belly"

serratus anterior: *serra*, "saw," and *ante*, "before"

transversus abdominis: *transversus*, "athwart" or "across," and *abdomen*, "belly"

NECK
scalenus: Greek *skalénós*, "unequal"

semispinalis: *semi*, "half," and *spinae*, "spine"

splenius: Greek *spléníon*, "plaster, patch"

sternocleidomastoideus: Greek *stérnon*, "chest," Greek *kleís*, "key," and Greek *mastoeidés*, "breastlike"

BACK
erector spinae: *erectus*, "straight," and *spinae*, "spine"

latissimus dorsi: *latus*, "wide," and *dorsum*, "back"

multifidus spinae: *multifid*, "to cut into divisions," and *spinae*, "spine"

quadratus lumborum: *quadratus*, "square, rectangular," and *lumbus*, "loin"

rhomboideus: Greek *rhembesthai*, "to spin"

trapezius: Greek *trapezion*, "small table"

SHOULDERS
deltoideus anterior: Greek *deltoeidés*, "delta-shaped" (i.e, triangular), and *ante*, "before"

deltoideus medial: Greek *deltoeidés*, "delta-shaped" (i.e, triangular), and *medialis*, "middle"

deltoideus posterior: Greek *deltoeidés*, "delta-shaped" (i.e, triangular), and *posterus*, "behind"

infraspinatus: *infra*, "under," and *spinae*, "spine"

levator scapulae: *levare*, "to raise," and *scapulae*, "shoulder [blades]"

subscapularis: *sub*, "below," and *scapulae*, "shoulder [blades]"

supraspinatus: *supra*, "above," and *spinae*, "spine"

teres (major and minor): *teres*, "rounded"

UPPER ARM
biceps brachii: *biceps*, "two-headed," and *brachium*, "arm"

brachialis: *brachium*, "arm"

triceps brachii: *triceps*, "three-headed," and *brachium*, "arm"

LOWER ARM
anconeus: Greek *anconad*, "elbow"

brachioradialis: *brachium*, "arm," and *radius*, "spoke"

extensor carpi radialis: *extendere*, "to extend," Greek *karpós*, "wrist," and *radius*, "spoke"

extensor digitorum: *extendere*, "to extend," and *digitus*, "finger, toe"

flexor carpi pollicis longus: *flectere*, "to bend," Greek *karpós*, "wrist," *pollicis*, "thumb," and *longus*, "long"

flexor carpi radialis: *flectere*, "to bend," Greek *karpós*, "wrist," and *radius*, "spoke"

flexor carpi ulnaris: *flectere*, "to bend," Greek *karpós*, "wrist," and *ulnaris*, "forearm"

flexor digitorum: *flectere*, "to bend," and *digitus*, "finger, toe"

palmaris longus: *palmaris*, "palm," and *longus*, "long"

pronator teres: *pronate*, "to rotate," and *teres*, "rounded"

HIPS
gemellus (inferior and superior): *geminus*, "twin"

gluteus maximus: Greek *gloutós*, "rump," and *maximus*, "largest"

gluteus medius: Greek *gloutós*, "rump," and *medialis*, "middle"

gluteus minimus: Greek *gloutós*, "rump," and *minimus*, "smallest"

iliacus: *ilium*, "groin"

iliopsoas: *ilium*, "groin," and Greek *psoa*, "groin muscle"

obturator externus: *obturare*, "to block," and *externus*, "outward"

obturator internus: *obturare*, "to block," and *internus*, "within"

pectineus: *pectin*, "comb"

piriformis: *pirum*, "pear," and *forma*, "shape"

quadratus femoris: *quadratus*, "square, rectangular," and *femur*, "thigh"

UPPER LEG
adductor longus: *adducere*, "to contract," and *longus*, "long"

adductor magnus: *adducere*, "to contract," and *magnus*, "major"

biceps femoris: *biceps*, "two-headed," and *femur*, "thigh"

gracilis: *gracilis*, "slim, slender"

rectus femoris: *rego*, "straight, upright," and *femur*, "thigh"

sartorius: *sarcio*, "to patch" or "to repair"

semimembranosus: *semi*, "half," and *membrum*, "limb"

semitendinosus: *semi*, "half," and *tendo*, "tendon"

tensor fasciae latae: *tenere*, "to stretch," *fasciae*, "band," and *latae*, "laid down"

vastus intermedius: *vastus*, "immense, huge," and *intermedius*, "between"

vastus lateralis: *vastus*, "immense, huge," and *lateralis*, "side"

vastus medialis: *vastus*, "immense, huge," and *medialis*, "middle"

LOWER LEG
adductor digiti minimi: *adducere*, "to contract," *digitus*, "finger, toe," and *minimum* "smallest"

adductor hallucis: *adducere*, "to contract," and *hallex*, "big toe"

extensor digitorum: *extendere*, "to extend," and *digitus*, "finger, toe"

extensor hallucis: *extendere*, "to extend," and *hallex*, "big toe"

flexor digitorum: *flectere*, "to bend," and *digitus*, "finger, toe"

flexor hallucis: *flectere*, "to bend," and *hallex*, "big toe"

gastrocnemius: Greek *gastroknémía*, "calf [of the leg]"

peroneus: *peronei*, "of the fibula"

plantaris: *planta*, "the sole"

soleus: *solea*, "sandal"

tibialis anterior: *tibia*, "reed pipe," and *ante*, "before"

tibialis posterior: *tibia*, "reed pipe," and *posterus*, "behind"

trochlea tali: *trochleae*, "a pulley-shaped structure," and *talus*, "lower portion of ankle joint"

CREDITS & ACKNOWLEDGMENTS

PHOTOGRAPHY

Photography by FineArtsPhotoGroup.com
Models: TJ Fink (tjfink@gmail.com) and Jenna Franciosa

ILLUSTRATIONS

All illustrations by Hector Aiza/3D Labz Animation India, except the insets on pages 10, 17, 18, 19, 20, 21, 22, 23, 24, 25, 28, 29, 30, 31, 33, 35, 37, 39, 41, 43, 45, 47, 49, 53, 55, 57, 59, 61, 63, 65, 67, 69, 71, 73, 75, 76, 79, 81, 85, 87, 91, 93, 95, 97, 99, 103, 104, 109, 111, 113, 115, 117, 121, 125, 127, 131, 137, and 141, and the full-body anatomy art works on pages 12 and 13: by Linda Bucklin/Shutterstock.

ACKNOWLEDGMENTS

The author and publisher also offer thanks to those closely involved in the creation of this book: Moseley Road president Sean Moore, general manager Karen Prince, art director Tina Vaughan, editorial director Damien Moore, designer and production director Adam Moore; and editors David and Sylvia Tombesi-Walton.

ABOUT THE AUTHOR

Hollis Lance Liebman has been a fitness magazine editor, national bodybuilding champion, and author. He is a published physique photographer and has served as a bodybuilding and fitness competition judge. Currently a Los Angeles resident, Hollis has worked with some of Hollywood's elite, earning rave reviews. Visit his Web site, www.holliswashere.com, for fitness tips and complete training programs. This is his third book.

AUTHOR'S DEDICATION

I dedicate this book to the love of my life—my fiancée Stacey Lynn Witner. She is the backbone of our family and the embodiment of strength, inspiring me in my own quest for knowledge and personal growth.